Managing Energy From the Top Down:

Connecting Industrial Energy Efficiency to Business Performance

Managing Energy From the Top Down:

Connecting Industrial Energy Efficiency to Business Performance

Christopher Russell, C.E.M.

CRC Press is an imprint of the
Taylor & Francis Group, an **informa** business

First published 2010 by The Fairmont Press, Inc.

Published 2019 by CRC Press
Taylor & Francis Group
6000 Broken Sound Parkway NW, Suite 300
Boca Raton, FL 33487-2742

First issued in paperback 2019

ISBN 13: 978-1-138-11266-7 (pbk)
ISBN 13: 978-1-4398-2996-7 (hbk)

Visit the Taylor & Francis Web site at
http://www.taylorandfrancis.com

and the CRC Press Web site at
http://www.crcpress.com

Library of Congress Cataloging-in-Publication Data

Russell, Christopher, 1962-
Managing energy from the top down : connecting industrial energy efficiency to business performance / Christopher Russell.
p. cm.
Includes index.
ISBN-10: 0-88173-625-2 (alk. paper)
ISBN-10: 0-88173-626-0 (electronic)
ISBN-13: 978-1-4398-2996-7 (Taylor & Francis distribution : alk. paper)
1. Industries--Energy conservation. 2. Industries--Energy consumption--Management. 3. Management--Environmental aspects. I. Title.

TJ163.3.R87 2010
658.2'6--dc22

2009039127

Managing energy from the top down : connecting industrial energy efficiency to business performance / Christopher Russell

Contents

Acknowledgements

Professional disciplines are sustained by their quality of mentoring. I hope to promote industrial energy efficiency by passing to young professionals the same quality mentoring that I received.

My greatest appreciation goes to the people whose patience I have tested most severely: Mike Pappas of Modular Process Control, Don Wulfinghoff of WESINC and author of the *Energy Efficiency Manual,* Bob Griffin of Enbridge Gas Distribution, Fred Schoeneborn of FCS Consulting, Jim Moore of TA Engineering, Christina Mudd of Exeter Associates, Gary Melickian (retired), Scott Rouse of Energy@Work, Bill Canis of the National Association of Manufacturers' Manufacturing Institute, Jim Eggebrecht of Texas A&M's Energy Systems Lab, Amy Wortman of the Energy Design Group, John N. "Skip" Trimble of South River Consulting, Kelly Paffel of Plant Support Engineering, Jim Pease of Unilever HPC, R. Neal Elliott III of the American Council for an Energy Efficient Economy, Anthony Wright of the Oak Ridge National Laboratory, Steve Thomas of Johnson Controls, Steve Heins of Orion Energy, Charles Rampey of Palmetto Energy Solutions, and Tommy Knight of ITA Energy Savings.

Much of the material presented in this book was developed from 1999 to 2006 when I served as the Director of Industrial Programs at the Alliance to Save Energy (www.ase.org), a Washington, D.C-based non-profit organization that advocates energy efficiency. Our industrial activities were funded primarily by the U.S. Dept. of Energy's Industrial Technologies Program.

While DOE-ITP is an outstanding source for technical energy information, the U.S. Environmental Protection Agency's Energy Star program for industry (www.energystar.gov/industry) may be the best public authority on the managerial and behavioral aspects of industrial energy use. There is also a vast network of state agencies, utilities, universities, trade associations, solution suppliers, and industrial energy users that devote time and resources to the promotion of energy efficiency. Through these organizations, I had the pleasure of meeting and working with MANY individuals whose imprint will be found throughout this book. Links to hundreds of other valuable information resources are located at www.energypathfinder.com.

It is appropriate to thank my parents, Colin and Alva Russell, for the head start. Last, but not least, thanks go to my wife Mary Ellen and daughter Alice for their love and uncomplaining support.

Christopher Russell
March 2009

Introduction

DEMOCRATIZING INDUSTRIAL ENERGY

In the U.S., over 200,000 facilities—highly varied in size, design, and purpose—manufacture products for delivery to the global economy. In the 21st century, one of the most pervasive challenges facing these facilities is the increasing cost of energy use. By simply opening its doors to operate, a manufacturer commits to using fuels and power, and is therefore exposed to the costs, liabilities, and public scrutiny related to energy consumption.

Solutions to these challenges are constantly evolving. Advanced technologies allow more work to be accomplished with less energy. Other technologies provide unprecedented control over existing energy-hungry equipment. Some energy waste can be avoided simply by changing the way work is scheduled or the way existing equipment is maintained. Some organizations eagerly adopt these changes. Others choose not to reduce the costs and risks that accompany energy waste.

Industrial energy waste persists for many reasons, both human and technical. In today's era of global competitiveness and scarce resources, "energy" is simply another factor that competes with safety, product quality, human resource constraints, environmental liabilities, and other concerns to receive management attention. All of these issues demand time and impose overhead costs, all of which divert resources away from the core business of meeting production goals. Organizations find it impossible to optimize all these agendas at once. Safety and liability issues usually come first—if only because of the legal dimensions of non-compliance surrounding these issues. And while energy waste will inflate costs, at least it poses no legal penalties (aside from those related to fossil fuel emissions). Accordingly, the energy agenda is forced to wait. A more provocative explanation points to the lack of incentives to achieve energy cost control. In other words, money is easily wasted on energy when there are no consequences for this outcome—even when key individuals are aware of the waste.

At the root of this dilemma is an imbalance of accountability within industrial organizations. Most other business issues can be neatly delegated to one department, or even one person, who then pursues the issue while everyone else carries on business as usual. Energy costs, however, reflect decisions made by virtually every employee in the facility. Choices—made one minute ago, or a month ago, or even 20 years ago—all play a role in shaping the current utility bill. Who is accountable for energy expense? A procurement director may be tasked with managing the price at which energy is purchased, but price is only one side of the expense equation. Energy consumption is the other side—and everyone from the receptionist to the chief engineer to the apprentice machine operator determines consumption patterns. These people never see the facility's utility bill, while the accounts payable clerk has no idea where the purchased energy went. Most facilities lack the ability to trace energy consumption at a meaningful level of detail. With no measurement, there is no accountability. Regardless of the price at which it was purchased, energy is effectively "free" to all staff. There is no compelling reason to economize the use of a free commodity.

Sure, corporate leaders want to do something about rising energy costs, but they don't have time for it. They have a business to run. Their reaction is to do what top managers are supposed to do: they delegate the issue to a subordinate. And the energy issue keeps getting delegated downward until it falls in the hands of someone who does have time for it (or who has no opportunity to delegate it any further).The lower you delegate energy cost-control duties, the more localized and temporary the solution. Typically, energy matters wind up in the hands a facilities person who lacks the authority to change the way that the wider organization uses, wastes, and thinks about energy.

Energy is now a risk management issue, especially in the wake of energy supply shocks that periodically shake the global economy. A growing awareness of climate change drives the rapid evolution of clean and efficient technologies and emissions-related regulation. In addition, information technologies increasingly support management decision-making on every corporate dimension, including energy management. What we have not seen is a concurrent evolution in the way industrial organizations respond to energy issues. Despite the onset of energy-related risk, most organizations still delegate energy issues to a mechanical department—usually adept at equipment maintenance, but often challenged to conduct the analytical and communication aspects of modern risk management. More importantly, delegation of this sort fails to engage the rest of the industrial organization, whose long-standing objectives, habits, and culture shape the company's energy expense. Effective responses to industrial energy risk will require cross-departmental collaboration and accountabilities that are largely without precedent.

Accordingly, the management of energy through the use of goals, behaviors, procedures, and accountabilities remains a vastly underappreciated business opportunity. While a few companies have developed highly sophisticated energy management disciplines, most organizations continue to underachieve in this regard. Unfortunately, the run-up of energy prices during the first decade of the 21st century comes on the heels of a decade during which manufacturing payrolls were slashed in response to global competitive pressures. The industrial tendency to wholly delegate energy issues to a "facilities" or similarly mechanical department is complicated partially by the chronic labor shortages that such departments experience.

A search for books on industrial energy management will reveal many publications, virtually all of which are written by engineers for a technical audience. These books tend to provide a hands-on, mechanical discussion of industrial energy systems and how to optimize their energy consumption. Most of these books are quite good and are worth having. However, books that address the organizational or "human" aspects of energy use are remarkably scarce. This non-technical book was written purposely to fill this void. Accordingly, this book will:

- inform industrial decision-makers about the business implications of energy use and waste;
- describe the potential for energy management as a means for controlling energy costs and building shareholder wealth;
- dispel misperceptions about energy efficiency and its outcomes; and
- demonstrate how energy management is developed and justified as an integral part of business strategy.

The intent of this book is to "democratize" the subject of industrial energy use. While technical people will continue to play a pivotal role in energy solutions, so will a number of other non-technical decision-makers who, prior to now, never imagined that they had a stake in energy cost control. Readers at all levels—from corporate leaders to machine operators—should take away a greater understanding of how wealth is harvested from industrial energy consumption.

Each chapter contains a series of short, easily readable sections. The overall flow is as follows:

- Chapter 1 explains the nature, scope, and magnitude of industrial energy use (and waste).
- Chapter 2 explains the business risks imposed by energy-related decisions.
- Chapter 3 examines the barriers to energy cost control.
- Chapter 4 reviews the organizational changes that pave the way for successful energy management.
- Chapter 5 examines the elements from which a company can build a strategy to manage energy on its own terms.
- Chapter 6 discusses "how the money works" in a way that purposely upsets the business-as-usual approach, offering instead a more robust method for financially justifying energy improvements.
- Chapter 7 offers some closing perspectives on the philosophy of energy management.

Appendices I and II offer discussions of energy procurement and electricity deregulation, subjects that are arguably outside the scope of this book, yet still of interest to many readers.

Chapter 1

Background

"The beginning of knowledge is
the discovery of something
we do not understand."
—Frank Herbert
(1920-1986)
U.S. author of "Dune" and its sequels.

Herbert's writing investigated ecology, leadership, and human evolution in general, thus evolving the science fiction novel beyond its dependence on some technological gimmick as its basis.

Money to Burn: Why Manufacturing Profits Go Up in Smoke

This first section is fiction in the truest sense. "Castiform Plastics" and the characters involved are all imaginary; any similarity to real companies or persons is coincidental. But while the fabric is fiction, the threads come from true stories—anecdotes collected from a decade of conferences, workshops, and plant visits. Understandably, no industry professional wants unflattering comments about his facilities to be published. That does not diminish the learning value that comes from observing others' mistakes. The intent of this prologue is to illustrate some of the very human reasons why so many industrial organizations simply fail to control their energy costs.

In a corporate boardroom high above the Dallas skyline, a sharp-penciled financial analyst projected a PowerPoint slide that furrowed the brows of the executives seated before her. This slide showed declining per-unit operating profits for the seven facilities that made up the Castiform Plastics business unit. Castiform was one of many diverse manufacturing enterprises owned by this publicly traded holding company. While the intricacies of plastics manufacturing and the specific causes of profit performance were of little concern to the corporate office, their quarterly earnings targets were. To stay in the good graces of Wall Street analysts, something had to be done, and quickly. Immediately after this meeting, a corporate directive to Castiform's general managers demanded that the declining trend in operating profits per unit be reversed.

A review of financials at the facility level revealed utility costs to be among the most rapidly growing expenses. "Utilities," a line-item expense that includes the water, fuels and power that plants consume to do manufacturing work, were universally perceived as a "technical" issue that demanded a "technical" solution.

At each Castiform plant, a general manager, or "GM" for short, hand-picked someone who was thought to be best equipped to investigate and solve the problem. At the Riverdale site, the GM wished to address the issue quickly and without diverting key supervisors' attention away from all-important production targets. Accordingly, the challenge was delegated to an enthusiastic young engineer who we'll name "Doug"—the kind of whiz-kid who tossed around technical acronyms and jargon with an ease that impressed other people, even if they didn't always understand what he said.

As a classically trained engineer, Doug's approach instinctively relied upon the unassailable principles of mathematics, physics, and thermodynamics. The initial scope of his inquiry therefore focused on plant utilities that converted fuel to heat and motive power. He identified discrete equipment—such as boilers and air compressors—as candidates for replacement with newer and presumably more efficient equipment.

Doug brought this to the attention of the plant's finance director, who expressed a general desire to devote capital to its highest and best uses. Energy systems were not considered core assets. To minimize investment risk for their scarce capital, the finance director would arbitrarily slash by 50 percent the estimated cash flow for any proposal that did not support core business activities. This move provided a margin of confidence that only the strongest proposals would receive capital funding support. On top of this, it would be eight months until the next round of capital budget development, and that was too long for Doug to wait.

A scan of internet resources allowed Doug to discover data-driven best-practice and benchmarking techniques for energy systems. This approach would, in so many words, add energy-saving practices to the everyday roles of machine operators and maintenance personnel. Some recommendations, such as "turn off equipment when not in use," were remarkably simple. After casually sharing this idea with colleagues over lunch, Doug gathered that the long-standing custom at Riverdale was to leave machines running even when no work was in process. It was felt (although not proven) that frequent stops and starts encouraged breakdowns, which would require writing up a maintenance ticket and losing precious time while waiting for the repair. Meanwhile, some grey-beard supervisors openly snickered at the thought of having some "college boy" tell them how to run their machines.

Similarly, Doug found a body of non-commercial guidelines for developing corporate energy management strategies. This material encouraged corporate commitment to continuous energy improvement. This involved top-down goal setting, planning, evaluation, reporting, and reassessment. While this concept immediately made perfect sense to Doug, he knew that its implementation would have to be championed at the general manager level. Doug was expected to solve the problem, not make additional work for his superiors.

Doug's next stop was Riverdale's procurement director, requesting a purchase order for securing an energy audit, which is an exercise that documents the opportunities to reduce a facility's energy waste. Unfortunately, the procurement director was not familiar with this concept, and was disturbed particularly by the term "audit." The Riverdale facility had recently endured a tax audit that resulted in a number of pecuniary penalties. Despite Doug's explanation, the procurement director was wary of the implications of such a study's findings. The director said he would approve the purchase of an energy audit only if the provider would tell him in advance what the energy savings would be.

After a month had passed, Doug recounted his findings for the GM. The ever-resourceful GM pondered the facts for a moment and had a revelation. The corporate directive required a restoration of per-unit profit margins. Nothing was said specifically about energy costs. The GM dismissed Doug from his office with a cursory thanks.

Doug walked away from his study of industrial energy with a striking revelation of his own: the problem is not *how much* money you burn—it's *when you burn it,* and from whose budget.

Given Doug's report, the GM privately came to an additional conclusion: with less than a year to go in his stint as Riverdale's top manager, he could effectively skirt the issue of higher energy costs by accelerating production to build up inventory. With machines running all the time, the trick was to maximize the volume of works in process, which would effectively lower expenses per unit of production. This would provide, at least for the short-term, a needed boost to operating margins that would satisfy corporate observers. Ramping up production also had costs—such as increased borrowing for raw materials and the additional inventory that would eventually weigh down the balance sheet. But those problems would not manifest immediately. This strategy allowed the GM to respond to short-term profit directives while buying the time he needed to run out his tenure. Debt and inventory issues would be problems for his successor to solve.

The GM reflected on a principle that had served him well over the years: you get promoted by solving today's problems, not tomorrow's.

ଓଃ

Short-term goals. Biased investment criteria. Localized accountabilities and turf issues. Timing of cash flows. Even if you read no further, you will understand that these are common organizational barriers to industrial energy cost control.

As you continue through the rest of this book, you will develop a greater understanding of the problems—and solutions—that characterize the business aspects of industrial energy use.

Energy is Wealth

Fuel and electric power are simply forms of currency that embody wealth. Currency creates wealth if it is used properly. Energy, like currency, is expended by a business to create new value. A well-run business controls its currency and the wealth it represents. That "currency" can and should include energy.

Consider how currency is used in business. Every business transaction represents some exchange of wealth. Sales, receivables, operations, payables, bank deposits, dividends, and investments—all of these represent value that is contractually recorded and carefully tracked according to well-documented accounting procedures.

Think now about energy, which is expended at every step in a business operation. Wealth, in the form of currency, is used to purchase fuel and power. In factories, energy purchases are converted to heat, pressure, and motive power that transform inputs into finished products. Those products are sold to generate income, which accumulates again as wealth. In commercial and institutional settings, energy improves indoor climates so that staff and clientele are made comfortable and productive—to create wealth.

Many organizations still perceive energy as something less than wealth. Facilities can lose valuable energy in a myriad of ways: inefficient combustion, steam and compressed air leaks, radiant heat transfer, and motor drives left running when there's no material in production. According to the U.S. Department of Energy, about 40 percent of the energy delivered to manufacturing facilities is lost to such insidious waste.[1] **The latest, most efficient technology in the world can still waste money if it is not maintained properly or if it is allowed to run unnecessarily.** The same U.S. DOE studies also show that up to half of that waste is economically recoverable. If maximum return on investment is a business goal, energy efficiency becomes a tool at management's disposal. In other words, energy efficiency ensures that energy dollars don't literally dissipate into thin air.

If manufacturers handle energy as anything less than wealth, there is no accounting of energy's contribution to value added. After it is wasted, no one tabulates the value of potential wealth that is lost. When companies direct their revenue into buying more energy to replace what has been lost, they direct money away from income that would have created more wealth. A large organization may have an army of clerks who track $20 taxicab receipts, yet no one can clearly account for the disposition of its $20 million in annual energy expenditures.

Think of all the management safeguards that an organization places on the currency and accounts in its control. Do fuels and power not deserve similar care? Energy management involves benchmarking of energy use, goals for improvement, metrics for ongoing measurement, and accountability for getting things done.

Why manage energy? Volatile fuel prices are only one reason. Energy use is pervasive throughout a facility, at every stage of production. If you monitor energy use, you have a pulse on the tempo of activity throughout the facility. If you pursue energy management, you develop leadership that can be leveraged for managing raw materials, labor, production schedules and other activities. In sum, these are the pivotal elements of industrial competitiveness.

Energy management establishes the means for guarding the wealth that energy represents. Energy, like currency, can be invested, preserved, and positioned to grow a business. Companies that understand this will manage energy like cash, tracking what they consume, waste, recover, and apply to the creation of wealth.

How Does Energy Use Affect Business Performance?

Energy is an ingredient common to all manufactured goods. Fuel is combusted to make heat and power, which are then used to transform raw materials into finished products. Energy transformation is sometimes messy and even dangerous, yet it is an inescapable aspect of industrial work.

The technologies, practices, and standards for optimizing energy transformation are known collectively as "energy efficiency." Energy efficiency initiatives are selected for their potential to reduce waste, minimize expenses, save time, reduce operating risk, and increase revenues through enhanced productivity. As an ongoing process, "energy management" mobilizes energy efficiency initiatives that improve business performance.

Energy management includes the monitoring, measuring, verification, and remediation of energy flows throughout a facility. Disciplined energy use results in control over production assets and the heat and power that they utilize. Reliable plant operations are the result of this control. Reliable operations are, by nature, more predictable in terms of on-time performance and workplace safety.

The money impacts of efficiency are several-fold:

- waste elimination reduces the volume of fuel and power to be purchased,
- reliable operations are predictable, which makes budget planning easier, and

- reliable facilities can fill orders faster, therefore filling more orders in the course of a year, thus generating more revenue.

Energy decisions have impacts beyond just the utility bills. A loss of plant reliability incurs extra costs in a number of ways. For one, the plant manager may be forced to incur overtime labor to compensate for downtime. To run these shifts, the plant manager may need to procure additional fuel and power on short notice. Customers tend to pay a premium for any short-notice purchase, and industrial plant managers are no exception. Energy efficiency can therefore help to reduce the average price of energy consumed as well as the amount of fuel required. Note also that interest costs always accrue, never waiting for idle machines. Add to this the revenue lost by not producing salable products.

This is a summary of how energy efficiency contributes to industry's needs:

- **Energy expense control.** One way to reduce energy expenses is to stop wasting fuel and power. The ability to manage energy use is a hedge against volatile energy prices.

- **Non-energy expense control.** Measures that improve energy efficiency also increase control over heat and power directed to process activities. Raw material waste is reduced when its manufacturing process applies heat at the right temperature, for the correct duration, in precise proportion to material inputs. Fewer wasted batches also mean fewer wasted labor hours.

- **Increased revenue potential.** When plant personnel boost energy efficiency through improved monitoring and maintenance of their energy-using equipment, they are also contributing to the mechanical integrity and reliability of that equipment. Productivity is enhanced with reliability. The facility's ability to generate revenue is directly related to its productivity.

- **Improved product marketing.** Companies are profiting by publicizing their resource stewardship efforts. Companies assure their buyers of the lowest prices by squeezing resource waste from the cost of their products. Like-minded businesses form alliances to serve the market for environmentally sustainable goods and services.

- **Speed.** In a global economy driven by cheap labor costs, American industry's competitive edge is predicated in part on speed. This means faster "cycle times" for process set-up and order fulfillment. Good energy management ensures the reliable delivery of heat and power throughout a production facility. Of course, plants with faster cycle times make more money.

- **Offsetting regulatory compliance risk.** In addition to the risk of operational failures, companies face many complex and changing liabilities related to environmental impacts. The combustion of fuel creates emissions that are subject to regulation. The reduction of fuel waste immediately reduces a facility's combustion emissions, therefore increasing the likelihood of regulatory compliance.

Energy efficiency's impacts are, in sum, consistent with the growth and preservation of shareholder wealth. In this regard, energy management bears a striking resemblance to financial planning. Decision-makers must:

- identify goals;
- select the investments needed to reach the goals;
- establish a blueprint and strategy for goal attainment;
- start early, if only with small efforts;
- maintain regular contributions over time;
- keep track of earnings; and
- manage risk through reinvestment and diversification of earnings.

Here, "diversification" means expanding beyond one-time energy projects to make energy management part of standard operating procedures throughout the organization. The financial and energy planning analogies share the same result—the growth and preservation of wealth.

Where Does Industrial Energy Go?

The industrial sector encompasses activities that extract and refine materials into the products that consumers eventually purchase. Certain basic materials such as fats and oils, pulp and paper, chemicals, petroleum distillates, and primary metals production are the building blocks from which final products are made. For example, an automobile is assembled from a vast array of intermediate materials, most of which were produced elsewhere before they were shipped to the point of final fabrication and assembly. The finished automobile contains material and sub-assemblies supplied by manufacturers of steel, aluminum, copper, glass, rubber, plastic, paint, and other intermediate products.

Most industrial energy is consumed by processes that yield bulk commodities to be delivered via pipelines, tankers, hopper cars, or barge loads to become inputs for other manufacturers. In turn, the goods that end-consumers ultimately buy are fabricated in batches by facilities that are, as a whole, far less energy intensive than the bulk process manufacturers.

A manufacturing corporation may operate one or more production sites. Each site usually features a number of structures that are arranged on a campus-like setting. The volume and form of energy required by a manufacturing site is highly varied—it depends on the type of product being made as well as the design of the production process and the structures that house those processes.

A distinction should be made between the total energy *de-*

livered to the fence of industrial facilities and the greater *primary* volume of energy that is ultimately earmarked for industrial consumption. Manufacturers have direct control over energy waste committed inside the fence. They cannot control waste committed in the generation and transmission of electricity *before* it reaches the fence.

Energy delivered to the fence consists of purchased electricity and fossil fuels such as natural gas, coal, or fuel oil. The purchased electricity, however, was generated off-site by some mix of coal, natural gas, nuclear, hydroelectric, and other processes. Of all the energy that goes into the generation and transmission of electricity, only about one third is delivered to consumers in the form of usable electricity.[2] The price for purchased electricity reflects these losses. This explains why a growing number of industrial facilities are (or are interested in) generating their own electricity *inside* the fence.

There are over 200,000 manufacturing facilities of all description in the U.S., providing about 14 million jobs while contributing $1.5 trillion to gross domestic product in 2005.[3] A typical industrial facility's layout and energy consumption may resemble Figure 1-1.

The entire site is a large property suitable for hosting production facilities, offices, warehousing and storage, employee parking, and infrastructure for moving materials on and off the property.

Energy delivered to the fence of industrial facilities is directed mostly to power houses, which in the U.S. as a whole consumed over 16 quadrillion Btu (~16 Q, or "quads") of fuel and power in 2002, down from 18 quads in 1998.[4] By converting fuel into heat and power, power houses represent the first step in organizing the energy needed to transform material inputs into a finished product. Material is transformed by adding and/or subtracting heat and by applying force and pressure. Energy is used for motive purposes, i.e., pumping, cutting, lifting, sifting, stirring, rotating, folding, etc., depending on the needs of the particular process. Electricity is used to operate

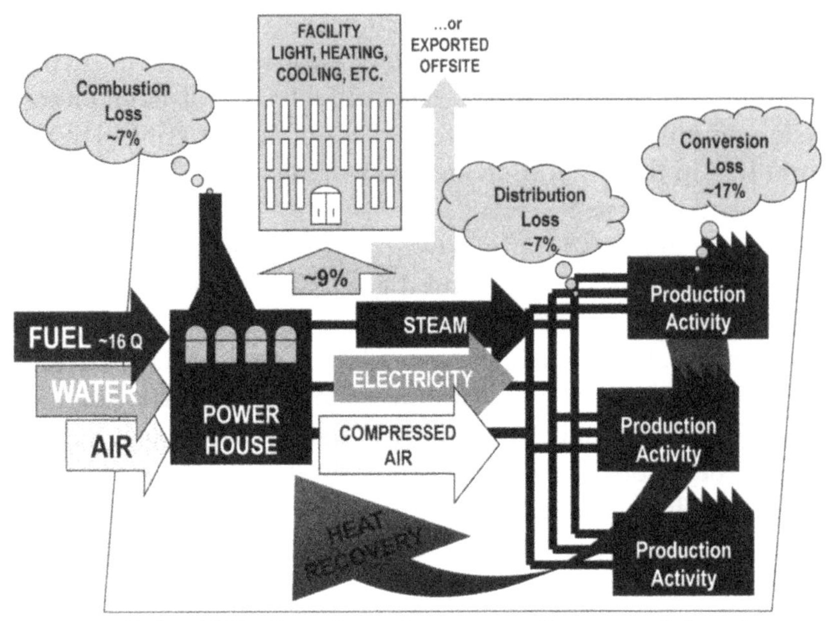

Figure 1-1. ***Source: Alliance to Save Energy***

a vast array of motor drives, fans, pumps, refrigeration units, lighting, air compressors, and incidental appliances. Compressed air performs a wide variety of repetitive motions. Also, energy is used to control the atmosphere in certain production facilities. The power house may host boilers, turbines, air compressors, and electricity transformers and switchgear. Boilers create steam, which is an efficient medium for transferring heat. Steam can also be used to power a turbine that generates electricity. When electricity is purchased from a merchant and/or utility, the power house merely transforms the purchased electricity into voltages required by the facility.

Note that industrial power houses experience energy losses as they convert fuel to heat and power. In the U.S., those losses are, on average, equivalent to seven percent of total industrial energy purchases.

The distribution infrastructure transmits steam, electricity, and compressed air from power houses to buildings that host production activities. Distribution system losses are characterized by leaks and radiant heat transfer, which on average add up to another seven percent of energy purchases. About nine percent of purchased energy is either exported offsite (in the form of excess electricity of steam from powerhouse activities) or is directed to human occupancy needs in the form of lighting, heating, cooling, and ventilation. Another 17 percent of purchased energy is lost as heat and power are converted to work in the production facilities. Some facilities recapture residual thermal energy from steam, hot water, and combustion emissions and return it for reuse by the power house. On average, only about 60 percent of industry's energy *received at the fence* performs production work, yet much of that waste can be economically recaptured.

Manufacturers' energy inputs typically follow this sequence:

- **primary energy input**, which represents fossil fuels consumed directly on-site by industrial facilities as well as the fuel used in generating electricity that is ultimately delivered to facilities;

- **central generation**, which mainly occurs in power houses where fuel is converted to heat and power by a steam plant, power generator or cogenerator (a single on-site process for generating heat and electricity);

- **distribution**, which carries heat and power from central generation to production areas;

- **energy conversion**, consisting of motors, fans, pumps and heat exchangers that transform heat and power to perform work; and

- **processes**, in which converted energy transforms inputs into final products.

The disposition of *primary* industrial energy, which includes the fuel inputs used for electricity generation, is summarized in Table 1-1.

Energy is lost at each stage of handling as described in Table 1-1. The fundamental laws of physics and thermodynamics make some losses unavoidable, but other losses are opportunities to embrace efficient technologies and practices.

How Waste Raises the "Price" of Energy

In reaction to rising energy costs, many business leaders will naturally focus on the price they pay for fuel and power. A "price-centric" approach seeks lower prices for the same fuel, or if possible, a switch to a different, lower-cost fuel. This is not a bad idea, but it recognizes only one side of the expense equation:

EXPENSE = PRICE times QUANTITY

Of course, any reduction in energy waste will reduce the total quantity of energy consumed. The companies that understand this concept proactively change the way they consume energy. Other companies—especially those that remain focused on prices—fail to grasp this opportunity. Price-minded managers should consider reducing their expenditure per unit of energy *available* to do useful work.

First, consider the relationship between fuel and the work it performs. Industrial facilities purchase fuel that must be converted several times in succession before it does the work for which it is intended. For example, consider steam systems, which consume over half of total fossil fuel purchases by industry. Almost all manufacturing processes require heat, and steam is an effective medium for heat supply. Fuel is transformed to heat in several stages, for example:

Table 1-1. U.S. Manufacturing Sector in 2002. Summary Allocation of Primary Energy Consumption

Stage of manufacturing energy use	*Volume of energy (trillion Btu)*	*Percent of primary energy*	*Characterization of losses*
Primary energy input	22,824	100%	
Offsite losses	-6,462		Energy is lost by power providers in the generation of electricity. Also, electricity and fuel lost is lost in transit to industrial facilities.
Power house (central plant) inputs	**16,362**	**72%**	
Steam & power generation loss	-1,226		Power house combustion efficiency determines the proportion of fuel that is converted to heat and power.
Energy distribution	**15,136**	**66%**	
Distribution loss	-1,177		Distribution pipes, vessels, and valves sustain a variety of leaks and radiation losses.
Energy for facility heat, cooling, etc.	-1,410		Not a "loss," but a reduction of energy available to process. These applications can also be inefficient.
Energy exported offsite	-86		In some states, manufacturers can sell surplus electricity that they generate onsite.
Energy conversion	**12,463**	**55%**	
Energy conversion inefficiencies	-2,838		A combination of inefficiencies, some avoidable and some not, are encountered as energy is converted to motive power used by motor drives, pumps, heat exchangers, etc.
Energy applied to process work	**9,625**	**42%**	An indeterminate volume of residual energy after process work is either reapplied to central generation or is lost without reclamation.

Source: U.S. Department of Energy. [4]

- **Fuel input to boilers**
  ...is combusted to:
- **Generate steam**
  ...which caries heat to a variety of:
- **Heat exchangers**
  ...which apply heat to transform materials.

Each stage of conversion allows some energy to be lost. The volume of loss depends on the quality of technology, procedures, and behavior applied by the facility and its staff. The U.S. Department of Energy's *Energy Use, Loss and Opportunities*[6] report describes the overall industry average energy losses incurred at each stage of the energy conversion process. While these figures vary across and within industries, it's useful to use aggregate industry measures:

- **Fuel combustion**
  ...experiences 8 percent fuel energy loss
- **Heat distribution**
  ...sustains 16 percent loss, and
- **Conversion of heat to work**
  ...sustains an additional loss of 16 percent

In other words, only about 60 percent of delivered energy is actually applied to the work for which it is intended. The other 40 percent includes waste that can potentially (and economically) be avoided.

So how does energy waste impact fuel prices? An example shows what happens when a facility purchases 100,000 units of natural gas (units are million Btu, or MMBtu) (see Table 1-2).

This facility acquired 100,000 units of energy at a price of $6.00 per MMBtu. However, after conversion losses, the $600,000 expenditure resulted in only 61,000 MMBtu available for actual use. Or, in other words, the facility spent $9.84 per *available* MMBtu.

Table 1-2. Illustration of Energy Value Lost by Hypothetical Facility, Using Industry Average Energy Conversion and Loss Rates

Sequence of activity	*Quantity (MMBtu)*	*Increment al value*	*Expenditure per "available" MMBtu*
Fuel delivered "to the fence" of the facility at $6.00 per MMBtu	**100,000**	**$600,000**	**$6.00**
Losses from combustion = 7%	-7,000	-$42,000	
Heat available for distribution	**93,000**	**$558,000**	**$6.45**
Losses from distribution = 7%	-7,000	-$42,000	
Energy for non-process use or export = 9%	-9,000	-$54,000	
Heat available for conversion to work	**77,000**	**$462,000**	**$7.79**
Losses from heat-to-work conversion = 16%	-16,000	-$96,000	
Energy available to perform intended work	**61,000**	**$366,000**	**$9.84**
Total energy losses = 39%	-39,000	-$234,000	

Another way to look at it: Energy changes hands several times after it is delivered to the facility. *At each step in the conversion sequence, the handler incurs energy waste that effectively marks up the "price" of the energy that is eventually applied to do useful work.*

Keep in mind that an energy system *feeds the leaks and losses first,* before it does the work it is intended to do. A facility's energy purchases and related infrastructure must be grossed up accordingly. For example, a facility may operate five air compressors where only three are needed, because no one is motivated to fix air leaks. By deciding to live with this waste, the facility commits to a greater investment (and carrying costs) in assets and equipment, excessive electricity purchases to run that equipment, and perhaps even the loss of valuable floor space that could be used more profitably. Energy-smart skills and procedures, developed prior to these capital investments, would reduce energy consumption as well as the cost of maintaining over-sized equipment.

The result shown in Table 1-2 is based on industry averages. Naturally, some facilities are better than others. But virtually all industrial facilities have the potential—through reduced energy waste—to improve their business performance.

Endnotes

1. United States. Department of Energy Industrial Technologies Program. *Energy Use and Loss Footprints.* January 5, 2009. http://www1.eere.energy.gov/industry/program_areas/footprints.html. Accessed March 2, 2009.
2. http://www.eia.doe.gov/emeu/aer/pdf/pages/sec8_3.pdf. Accessed March 2, 2009.
3. National Association of Manufacturers, *The Facts About Modern Manufacturing, Seventh Edition.*
4. United States. Department of Energy Industrial Technologies Program. *Energy Use and Loss Footprints.* http://www1.eere.energy.gov/industry/program_areas/footprints.html. Accessed March 2, 2009.
5. *ibid.*
6. United States. Department of Energy Industrial Technologies Program. *Energy Use and Loss Footprints.* http://www1.eere.energy.gov/industry/program_areas/footprints.html. Accessed March 2, 2009.

Chapter 2

What's at Stake?

"Not everything that is faced can be changed,
but nothing can be changed until it is faced."

Lucille Ball
(1911-1989)
Comedienne, model, actress, and film executive

Once told by drama coaches that she had "no future at all" as a performer, Lucille Ball methodically spring-boarded herself from modeling to Broadway, from B-films to television, and eventually to executive producer and cultural icon.

"We're Already as Efficient as We Can Be!"

An industrial facility has achieved truly efficient use of its energy if the following conditions are true:

- You know *exactly* how much energy you are using, right now, per unit of production, per square foot, or hour of operation, or some similar metric. You have a constant flow of data to prove it. The appropriate staff have a protocol for reacting when that energy performance data varies over time. Individuals are not just empowered to take action, they are trained in remedial methods and are held accountable for results. The remedies are documented and made available for future reference. Finally, all of your past improvements remain fully implemented. Checks and balances ensure that no one takes procedural shortcuts to circumvent these improvement measures.

- You can demonstrate the professional or industry standard by which you achieved your maximum efficiency. You monitor the evolution of these standards over time and update your facility procedures accordingly. You are aware of new technologies and are prepared to adopt these as cost-benefit criteria allow.

- You retain all staff with institutional knowledge of energy and utility systems. Or, you maintain strategic relationships with vendors, consulting engineers, or other partners who provide the expertise that is lacking in-house. All of your new hires are well-versed in energy-smart behavior, or at least you integrate this into training and orientation.

- Your energy performance metrics are continuously adjusted to account for the depreciation of assets through constant use.

- You thoroughly account for variability in fuel prices and interest rates over time. Also, you are instantly aware of various tax investment incentives are offered by state and federal government as well as your utility. The feasibility of energy improvement initiatives change with these variables. You watch these variables and are prepared to act quickly on a pre-existing list of potential energy improvements. Accordingly, your investment decisions are triggered by movement of these rates and incentives.

If all of these conditions are met, your facilities are truly as efficient as they can be.

Ask Not What *Can* You Save... But What *Will* You Save?

The question that industrial decision-makers will most frequently ask about energy cost control is "How much can I save?" This chapter provides some typical measures based on recent program experience. The question that they really should be asking is "How much am I likely to save?" An answer to this question will also be offered below.

WHAT CAN YOU SAVE?

The average industrial facility can expect to economically reduce its energy consumption somewhere within a range of 10 to 20 percent. Keep in mind that "10 to 20 percent" describes an average range of expectations. Some facilities can capture more savings, some less. If you want a more precise number, you will need to conduct an energy audit—a facility-wide study of energy inputs, uses, and losses (see page 71). Keep in mind that energy audits are a very human process, reflecting the skills and experience of the team that conducts them. Ten different audit teams can examine the same facility—and develop ten different sets of recommendations. Their findings may generally overlap, but each report will present different cost-benefit evaluations, suggested priorities, or even unique findings.

Industry's best options for reducing energy costs were summarized in a study sponsored by the U.S. Department of Energy. [1] In total, *economically achievable* energy-saving opportunities represent 5.2 quadrillion Btu—21 percent of primary energy consumed by the manufacturing sector. These savings equate to almost $19 billion for manufacturers, based on 2004 energy prices and consumption volumes.

10.4 Percent?

Almost 2,000 industrial energy assessments were compiled by the U.S. DOE-sponsored *Save Energy Now* program from late 2005 through 2008.[2] Although purposely limited in scope to focus on easy, quick-payback savings opportunities, a preliminary summary[3] of results published in 2007 found that these assessments identified an average of 10.4 percent annual energy savings potential per facility. The average dollar value of annual savings identified per facility was over $2.5 million. By the end of 2008, assessments had been conducted in 46 different states. Most facilities were in chemical manufacturing, paper manufacturing, primary metals, food, non-metallic mineral products, and fabricated metal products industries.

>10-15 Percent?

See the U.S. Department of Energy fact sheet entitled *Save Energy Now in Your Motor Systems*.[4] It includes comments about all potential sources of industrial energy savings, not just motors. According to this document, plants with an energy management program already in place can save an additional 10-15 percent by using best practices as recommended by the U.S. Department of Energy. Remember, that's *in addition* to an existing energy management program.

23 Percent?

Sixty-six steam system audits in Ontario, Canada were compiled between 1997 and 2002 by Enbridge Gas Distribution through its "Steam Saver" energy audit program.[5] Participating facilities included a wide variety of manufacturing and large institutional facilities, with annual gas bills ranging from tens of thousands to tens of millions of dollars. The average identified savings potential (weighted for the volume of each plant's annual energy consumption) was 13.2 percent. The arithmetic average of individual plant savings was 23 percent—this means that smaller plants showed larger percentage savings. The distribution of the Steam Saver's first 66 plant audit results is shown in Figure 2-1.

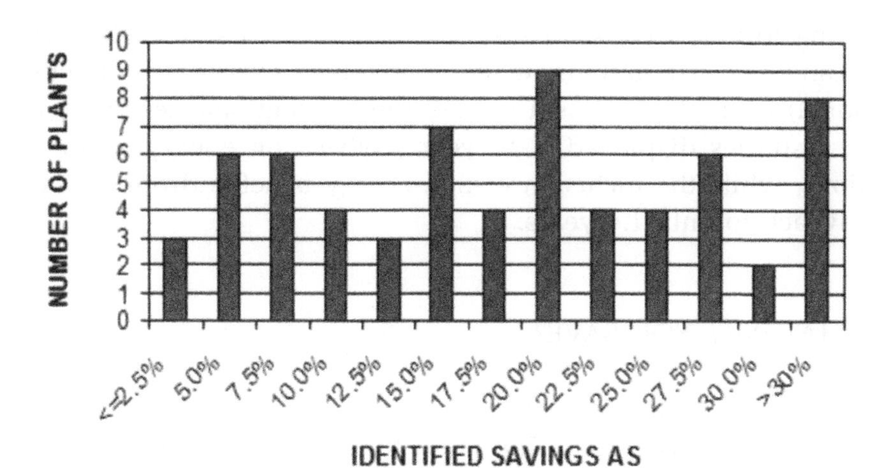

Figure 2-1. Distribution of Identified Annual Gas Savings Potential, 66 Industrial & Commercial Steam Plants, Enbridge Gas Distribution "Steam Saver" Program, 1997-2002

24 Percent?

See *Energy Loss Reduction and Recovery in Industrial Energy Systems*.[6] This U.S. DOE document claims, on page 22, that industry's overall energy consumption can be reduced by 24 percent through efficient technologies and practices. Appendices in this report provide industry-specific claims for energy savings potential.

This cannot be overemphasized: no single industrial facility is "average." Each facility features a unique design, purpose, product mix, operating schedule, maintenance history, and work habits. Savings potential varies accordingly.

Good energy decisions can reduce certain non-energy expenses as well. A summary of 77 case studies gives some indication of the value of non-energy benefits attributable to energy efficiency in a manufacturing setting.[7] Of the total number of cases, 52 included a monetized estimate of both energy and non-energy savings. Based on energy savings alone, project paybacks in aggregate were 4.2 years. With non-energy benefits included, the aggregate payback was 1.9 years. It is also interesting to note that 41 of the 77 cases involved state-of-the-art technology installations, while 35 involved everyday (conventional) technologies. As a subset, the conventional technology case studies displayed a 2.3-year payback on energy savings alone, while the inclusion of non-energy benefits dropped the payback to only 1.4 years.

WHAT *WILL* YOU SAVE?

This is really the question that business leaders should ask. Consider this checklist. The more times you can answer with a "yes" to these questions, the more savings you are likely to achieve, and the faster you are likely to achieve them.

- Will you obtain an energy audit—a thorough, facility-wide study of energy inputs, uses, and losses? (See page 72)
- Will your staff know the purpose of this energy study and not be intimidated by it?
- Will your facility support energy cost control as an ongoing process rather than as a one-time project? (See page 85)
- Will top management stand behind the goals and accountabilities set by an energy management plan, and not ignore them after a year has passed?
- Will staff be responsive to energy awareness training?
- Will operations, maintenance, and procurement people be willing to change the way they do things by incorporating energy best practices into their work habits?

Take heart—no one answers "yes" to all these points. But as you achieve more "yes" answers, the more you are likely to save.

Business Risks for Energy Consumers

Any organization that consumes energy should be prepared to manage a wide variety of energy-related business risks. These include energy market volatility as well as rapidly evolving technologies and regulation. Dealing with these risks involves more than pursuing a "project"—such as capital investment in a big chunk of machinery. It may involve durable management strategies that change the way people in an organization use (and think about) energy.

Energy risk management will require input from a variety of departments and people throughout an organization. The management challenge is to "democratize" the energy management role so that all industrial stakeholders understand how they can contribute to—and benefit from—the wealth that their energy consumption represents:

Procurement, budgeting, and finance people will be the front line in dealing with energy prices and the challenges imposed by the deregulation of electric utilities. Companies need to develop strategies for making the best use of the many procurement options that are available in deregulated power markets.

Finance people will lead the pursuit of tax deductions and credits that apply to certain energy improvements such as lighting, heating, air conditioning, and building structural systems. They will also establish investment criteria that facilitate (or hinder) the deployment of efficient assets and equipment.

Engineers will monitor emerging technologies and standards. Companies will ask: What are these technologies? Which ones will provide value for us? How shall we evaluate them? Engineers will answer these questions and also design, commission and monitor new energy-using equipment and systems.

Operations managers will rethink the dozens of staff decisions made each day that impact energy use across plant floors or office spaces. Machine operators, production schedulers, and office workers are largely unaware of how their every-day choices impact the energy bill. Solutions begin with increased awareness.

Human resource professionals need to inventory their staff training needs, and then seek appropriate training opportunities. Maintenance workers and machine operators need to learn best-practice techniques that save energy and boost reliability.

Environmental, health and safety professionals need to monitor emerging regulations. Energy consumption directly determines the emissions output of a manufacturing facility. Compliance with these regulations puts many dollars at stake in the form of potential fines and penalties. Note that an energy management agenda will closely overlap environmental, health, and safety compliance strategies.

Marketing and corporate strategy people need to understand the opportunities posed by "sustainable" business practices. Energy efficiency is usually the first and largest component of sustainable business practices (see page 156). Sustainability is also the key to developing new products and services and winning new customers. Look at Wal-Mart: they force their suppliers to squeeze as much waste as possible from their production costs. Companies that sell their products to Wal-Mart (and many other like-minded firms) need to be aware of this trend and have a strategy ready for it. To ignore this trend is to risk losing business.

Needless to say, someone needs to coordinate these many players so that they are not working at cross-purposes. *This is essentially the role of an energy manager* (see page 112).

Forward-thinking companies respond to energy risk by changing the way they use energy. They often begin by reevaluating their work habits and procedures. They quickly discover that energy use is as much a human issue as it is mechanical. To ignore the human component of energy cost control is to invite business risk. A lack of awareness begets a lack of accountability. And without accountability, companies have no effective response to business risks imposed by energy use.

The Promise and Challenge of Life-cycle Cost

"Life-cycle cost" describes the total cost of ownership for any asset placed in service by a business organization. This is more than just the "catalogue" or acquisition price. There are search costs incurred prior to purchase, the costs of finance and installation, insurance, then the costs of maintenance, operating inputs (especially energy), and the asset's eventual disposal. Energy-saving industrial hardware is usually described in terms of the life-cycle cost benefits that it provides. These benefits are relevant to an organization that seeks to minimize its operating costs over time.

Here's the problem: organizations are made up of individuals with different responsibilities and sometimes very different time horizons, which means that some are looking for results today, while others are focused on results months or even years from now. Add to this the fact that individual objectives may vary across departments and facility boundaries. There's usually no one person who is responsible for maximizing value simultaneously across all these dimensions.

A good starting place for this discussion is the procurement director. First costs—meaning the initial cost of buying an asset—are the sole focus of most procurement directors. The procurement director's neck is on the line for minimizing costs today. In many cases, he couldn't care less if the lowest-cost assets actually lead to excessive energy or maintenance expenses for years to come. Those costs simply become an operating manager's problem. The procurement director makes his bonus today by keeping first costs to a minimum.

Similar constraints impact finance directors. Corporations usually impose annual or even quarterly earnings targets on production facilities. At the same time, production depends on assets that slowly wear out over time or are made obsolete by the advent of newer, more efficient alternatives. Finance directors are always tempted to coax more life out of old equipment rather than investing in upgrades. This approach will artificially boost earning for this quarter, and maybe the next. Finance directors can hit their short-term targets and run, while the problems associated with eventual asset failure and escalating operating costs will hopefully accrue later, to the next manager.

Most major assets have significant ancillary needs such as lubricants, tools, filters, and other consumables. These items are also subject to procurement from the lowest bidders—supposedly ensuring that the company's out-of-pocket expenditure is minimized. Unfortunately, the "cheapest" selection is not always the most durable. That which wears out faster is replaced more frequently. Another complication may be the structure of vendor relationships, which blend service with the supply of components dictated by the vendor.

Operations staff present a different perspective. These people are focused on production schedules. Time is a valuable commodity in operations, to the point where other factors, including energy, may be readily sacrificed in order to save time. For operations staff, job risk is related to production goals, not energy waste. Allowing machines to run without work in progress is of no consequence to machine operators who neither see nor pay the utility bills.

Organizations pay a premium—in the form of energy waste—for failing to minimize life cycle costs. This premium is value waiting to be captured. The advocate for this is an energy manager. In practice, the energy manager seeks a balance among financial, technical, and operational criteria. It should be clear now that energy cost control is as much people-oriented as it is technical. Authority, communication and persuasion are crucial to the task. Execution of this agenda becomes more urgent as the price of energy rises.

Embedded Energy

A major opportunity for manufacturers is to identify embedded energy costs. Managers at each stage of a manufacturing process may overlook energy waste because energy is only two, three or five percent of production costs. But the prices of final products must absorb layers of energy inputs as materials are progressively transformed into final products.

For example, the direct energy cost for assembling an appliance might be only a few dollars—and a very small fraction of its retail cost. But in the big picture, there was energy consumed in mining the appliance's iron ore, copper, and bauxite; in metal treating; in rubber and glass manufacture; in power house fuels for the facilities that make plastics, paints and dyes; and in energy feedstocks, which are energy commodities consumed directly as product ingredients. Any waste of energy in the manufacture of these intermediates, disguised in the cost of inputs, eats up profit margins at every step. In effect, consumers are "taxed" for any waste committed at all stages of the manufacturing process. This is true for appliances, consumer electronics, toys, processed food, or anything that is manufactured for purchase by the final consumer.

A product's energy footprint describes its total energy impact, including its design, manufacture, shipment, customer use, and its eventual disposal at the end of its operating life. Historically, if industry had any interest in energy consumption, it ended when products were finished and shipped. Today, however, consumers are increasingly concerned with the energy consumed by their appliances, cars and homes. For that reason, manufacturers should be, too.

Today's energy challenges present industry with an opportunity to think broadly about innovation—not only as it applies to production processes, but in the way products are developed and eventually used by consumers. Manufacturers can harvest energy-related wealth in two ways—externally, by better serving the market that buys their products, and internally, from operating efficiencies.

The energy footprint concept outlines the opportunities to create superior product value. The first stage minimizes energy waste during product manufacture, warehousing, distribution, and sale. The next stage represents potential energy savings in product use by the consumer. Finally, product design ensures that energy and environmental impacts are minimized by the disposal of the product at the end of its useful life.

Manufacturers can partner with their suppliers to map their energy intensity to strategically squeeze out avoidable costs. Technology research and development (R&D) is crucial, but so is parallel development of human skills to manage energy use by large organizations. Companies are always partnering to achieve economies in distribution and inventory, so why not in energy management? The information technology exists, and it can be done.

Beware of "Fugitive" Energy

Energy always performs work. The energy harnessed to do industrial tasks will either work *for* or *against* the manufacturer; but it never stops working. Most of the total energy obtained by a manufacturing facility (hopefully) ends up performing the work for which it is intended. The rest "escapes" to perform unintended work. The same characteristics that make energy valuable—as a provider of heat, pressure, and motive power—also make it potentially destructive. This premise allows us to introduce the concept of "fugitive energy."[8]

To understand this, consider the total volume of fuel and power purchased by a facility. From that delivered total, some fuel is converted to heat, and as heat and power are converted to work, some energy, like heat emanating from a smokestack, is simply released to the atmosphere. But energy is not only lost to thin air. Because it dissipates in the form of heat, friction, vibration, and chemical reactions, **energy also contributes to the destruction of the machines and fixtures through which it travels**. This is referred to here as "fugitive energy."

Examples of fugitive energy are many. They include the corrosion of the interior of metal smokestacks attributable to fossil fuel combustion gasses that condense to form acids. Undissolved gasses in boiler feed water promote similar corrosive effects inside steam distribution hardware. Energy is misapplied as "water hammer" when high-pressure steam collides with stagnant water in a distribution main. Water hammer can rip pipes and valves from their moorings with force that can injure or kill bystanders. Poor quality electric power can cause motor drives to overheat and fail prematurely.

Energy is wasted when machinery is left running while not actively producing products. Pumps rigged to run at full capacity over the weekend (so that maintenance crews won't have to get up during the football game) are sustaining additional friction that will shorten their operating life. Older-technology light fixtures expend unwanted heat while also causing air conditioning systems to work longer, consume more energy, and fail earlier.

A proper accounting of fugitive energy only begins with the value of wasted energy purchases. It also includes the premature depreciation of equipment, wasted material and works-in-process, settlement costs associated with personnel injuries, fines assessed when fossil fuel emission limits are exceeded, and the interest costs (and lost revenues) attributable to downtime. Consider also the floor space given up to equipment that is oversized or redundant. In many instances, the right-sizing of energy-using equipment will reduce energy costs and relinquish floor space to more productive uses. Floor space wasted this way is also a fugitive energy cost.

Industry is generally familiar with energy risk imposed by forces external to their facilities. These include volatile energy markets, inconsistent quality of energy commodities, discontinuous supply, and technological change. Contrast this to fugitive energy, which is a self-imposed risk suffered by facilities that delay, defer, or otherwise ignore potential energy improvements. Fugitive energy is the kind of energy risk that industry should be most able to control, or in other words, the dollar losses that should be easiest to avoid.

The Competitor Within

Manufacturing, along with mining and agriculture, is the true foundation of an economy. All other organizations—banks, realtors, wholesalers, retailers, software developers, places of worship, professional sport leagues, social clubs, and so on—merely rearrange the value that the industrial foundation produces.

As the level of manufacturing sector employment in the U.S. declines, so does the number of business leaders with technical capabilities. Science and engineering majors represent a diminishing percent of bachelors and masters degrees awarded by U.S. universities.[9] There is a growing disconnect between non-technical corporate leaders and the people who make things happen on the plant floor. Industrial investment decisions are increasingly made by people have no concept of how heat, force, and motive power are applied to raw materials to transform them into the products we use every day. Similarly, many policy leaders know nothing about industrial energy use beyond the fact that prices are too high.

Also, let's remember that our manufacturing facilities have fewer production people trying to cover more responsibilities. As a consequence, competition arises within industrial corporations, especially at budget time, pitting plant against plant, department against department. Example? During a recent plant walk-through, a question about a persistent steam leak was met with the response "it's not my leak."

There are no energy villains or culprits to apprehend. Energy waste is fostered simply by a lack of accountability to do anything about it. Consider Figure 2-2, which compares the decision-making environment of the homeowner to that of a typical manufacturing facility.

Residential energy decision-making resides solely with the homeowner. In a manufacturing facility, these decisions are distributed across a number of departments, dependent on individuals whose priorities usually outweigh concerns about energy costs. Without a protocol for coordinating roles, costs, and ben-

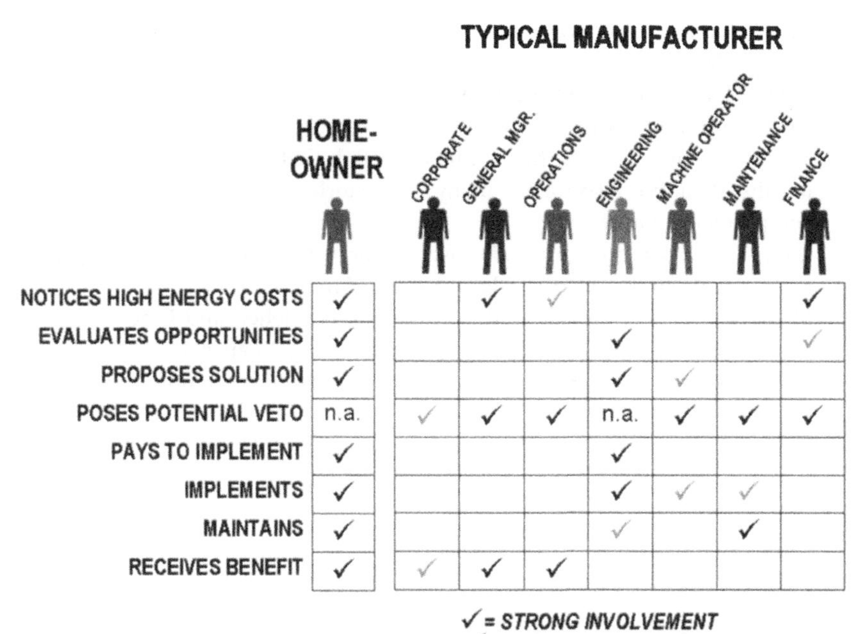

Figure 2-2

efits, energy initiatives will seem like a distraction to managers who are already overburdened with "normal" duties. In addition, managers will veto proposed energy improvements if these changes are perceived as a threat to their normal routine.

"Competitors" reside inside many facilities in the form of waste that is allowed to happen. This waste inflates the cost of the goods sold, effectively reducing profit margins. Awareness can keep this waste from growing; accountability can reduce its impact. Costs can be converted into profits.

Endnotes

1. United States. Department of Energy Industrial Technologies Program. *Energy Use and Loss Footprints.* http://www1.eere.energy.gov/industry/program_areas/footprints.html. Accessed March 2, 2009.
2. http://apps1.eere.energy.gov/industry/saveenergynow/partners/results.cfm. Accessed March 2, 2009.
3. http://apps1.eere.energy.gov/industry/saveenergynow/partners/pdfs/sen_2006_results_summary_9-20-07.pdf

4. http: / / www1.eere.energy.gov / industry / bestpractices / pdfs / 39157.pdf. Accessed March 2, 2009.
5. This has since been updated to at least 92 steam plants, as reflected in a summary report: http: / / www.steamingahead.org / library / enbridge05.pdf. Accessed March 2, 2009. Many thanks to Bob Griffin not only for sharing this data, but also for his many "war stories" from the front lines of energy management.
6. http: / / www.eere.energy.gov / industry / energy_systems / pdfs / energy_use_loss_opportunities_analysis.pdf. Accessed March 2, 2009.
7. H. Finman and J. Laitner. "Industry, Energy Efficiency, and Productivity Improvements." U.S. Environmental Agency White Paper. 2002.
8. This is the author's term for a concept brought to his attention by Bill Adams of Flowserve.
9. http: / / www.nsf.gov / statistics / nsf08321 / . Accessed March 2, 2007.

Chapter 3

Overcoming Barriers

"*If it ain't broke, don't fix it* is the slogan
of the complacent, the arrogant or the scared.
It's an excuse for inaction, a call to non-arms."
—Colin Powell
(1937-)
American statesman and retired four-star general of the United States Army.

Colin Powell rose in rank as a soldier and maintained his credibility as a statesman specifically because of his determined focus on principles as opposed to expedience.

Why Do We Resist Energy Efficiency?

Manufacturers are keenly aware of energy market volatility and its impact on their earnings. Most energy policy discussions merely call for relief in the form of increased energy supplies, to be achieved by stepping up power plant construction and through more drilling and digging of fossil fuels. Certainly, new energy supplies must be secured. But how long will it take for these new supplies to come to market? Five years? Or ten years? And how much new capacity should be added?

Keep in mind that of all energy delivered to U.S. industrial facilities, about 40 percent is not applied as intended to works in progress.[1] In other words, a lot of energy is wasted. Avoiding much of that waste is not only economically feasible at the facility level, but in fact desirable from the general public's perspective.

If energy consumption is inflated through waste, then investment in new power generation and refining facilities will be similarly inflated. Building new energy production capacity is fine, but let's minimize that investment so that wealth can be directed to more productive investments.

Energy waste happens because misperceptions surround not only the concept of energy efficiency, but also its anticipated impact on daily job functions. Many corporate leaders genuinely believe that energy use is an uncontrollable cost of doing business, so they focus solely on energy prices. Then there are business leaders who truly believe that their facilities have already eliminated energy waste. This conclusion is often based on bad information, including different understandings of what constitutes "energy efficiency." Some common misinterpretations include:

- installing a backup generator
- participation in a municipal recycling program
- switching fuels
- hedging of energy price risk with futures contracts
- purchasing energy through a non-utility marketer
- cutting back production or services to reduce energy consumption
- adopting renewable fuel sources, such as solar or wind power
- installing dual-fuel capabilities on boilers and other large energy-consuming assets

With misunderstanding comes resistance and some very large hurdles. For some very understandable reasons, industry's top managers are very reluctant to even admit to energy waste. Think about it: what reaction should be expected when a vice president of operations recognizes avoidable waste that has been sustained under his or her watch? A strategy of denial is sometimes the key to survival.

The classic engineering solution for controlling industrial energy costs is to install new equipment—a capital investment project of some kind (see page 85). This approach assumes that the organization will approve any project with a good payback. That's not always true. Organizational size and complexity often pose formidable barriers to making financially sound energy improvements (see page 59). Manufacturing enterprises have long-standing organizational structures, accountabilities, and incentives that are designed to make products and get them out the door, not to "save energy."

While most companies will express a desire to reduce costs, waste is not fully recognized in day-to-day practice. Also, facilities are organized into departments that typically compete with each other for budget resources. This internal competition is the underpinning of "turf" issues and a your-problem-not-mine mentality. Proposed energy improvements may be conceived in one department and developed and paid for by another, while the benefits may appear on yet another department's books (see Figure 2-2 and related discussion on page 32). Any number of department heads can veto the proposal, but typically, only one decision-maker can approve it. Control of energy waste requires cross-functional authority and communications that don't exist in most facilities. Organizational shortcomings, and not individual behavior, are at the root of energy waste. Given this reality, energy waste will continue, no matter how financially attractive a project looks on paper.

Industrial facilities that lack their own, internal energy cost-control mandate will allow waste to continue. Successful energy cost control requires cross-departmental planning, cooperation, and leadership (see page 112). Unfortunately, energy waste is an inexorable drain on industrial earnings. It does not wait for industry to figure out how to defeat it. At the same time, it inflates estimates for future investment in energy infrastructure—investments that could be directed to more productive uses, both from business and from societal perspectives.

How Did We Get Here, and How Do We Regain Control?

How does a manufacturing facility become less than energy-efficient? Much of it has to do with the degradation of production assets over time, coupled with the emergence of newer, more efficient technologies.

The reasons are also human. "That's the way we've always done it" is the justification for long-entrenched work habits that become default procedures. Historically, industrial organizations simply haven't had the need, motivation, or mechanisms to account for their energy use.

The energy used today by a certain industrial facility reflects many decisions made by many people over time. Fundamental decisions were made when the facility's process and supporting energy systems were designed. These decisions were long-term commitments to a certain vintage of technology. Large, fixed assets will operate for years or even decades at a time. This includes equipment such as boilers, furnaces, and air compressors. Complex production systems include a wide variety of smaller equipment such as pumps, fans, and motors. Compared with the larger assets, these smaller components are more easily replaced. However, the design of the overall system in which they were installed is not as easy to change. Industries typically conduct multi-year planning cycles to organize major facility upgrades and system changes. Planning cycles allow facility managers to avoid frequent disruptions to their production schedules. These cycles take years to conduct and involve a number of considerations, including energy costs. This partially explains why many manufacturers do not respond immediately to proposed energy improvements, even if investment incentives or rebates are involved.

Procedures for operations, procurement, and maintenance are also very durable, being subject to change on an annual basis or even longer. Many procedures are instituted without energy cost considerations in mind. Finally, energy consumption reflects the decisions made by equipment operators on a daily or even hourly basis.

Keep in mind that the people who design and select industrial equipment are usually quite different from the people who maintain and operate the equipment. Maintenance and operations personnel have a variety of responsibilities, needs, and motives for making the choices they do. These choices become institutionalized habits over time. Certain habits that save time and effort may be at the expense of excess energy consumption. These practices had little consequence when energy was cheap. However, the trade-offs between time and money change as energy prices escalate. Attempts to change these work habits can cause friction among staff. Energy improvement proposals that somehow threaten the status quo of procedures are quickly dismissed—if staff have no accountability for energy waste.

A manufacturer's decision to make energy improvements must compete with other priorities. Procurement officers are often compelled to make equipment purchases based on the lowest cost of acquisition, not the total cost of ownership. Long-term vendor/supplier agreements may make it easier to acquire the "wrong" kind of equipment. Production deadlines may force operations personnel to make emergency repairs using whatever equipment is available, as opposed to what is optimal from a total-cost-of-ownership perspective. Energy performance considerations are usually secondary to a plant manager's need to meet production targets.

When a facility fails to *manage energy as a form of wealth,* its staff begin to use energy in ways that engineers and other technicians never anticipated. These "services" represent industry's unintended demand for energy. Here are some examples:

- **Proof of effort**. In today's globally competitive economy, a worker's survival depends on keeping busy, or at least *appearing* to keep busy. This sometimes explains why many operators prefer to leave certain machines running, even when there is no work in process. Motor drives, pneumatic tools, and other factory machinery all make a distinct racket. A manager can, after a time, detect what machinery is running without having to look—it can simply be heard. The sound implies that "yes, we are busy." The energy wasted by machines that run unnecessarily is of no consequence to the worker—the energy cost is not reflected in his or her paycheck. However, energy provides a valuable service to the operator who wishes to maintain the appearance of keeping busy.

- **Budget defense**. Many organizations maintain the fiscal habit of developing next year's budget based strictly on the previous year's performance. In other words, the department that successfully decreases its expenditures this year can actually be penalized with a smaller budget next year (see more about this on page 55). While few managers actively promote waste, many more simply won't challenge it. They can then confidently prop up their funding request for the coming year's budget. In terms of energy use, this again means running machines unnecessarily, using fuel-rich combustion settings, and ignoring leaks and losses.

- **Comfort and convenience**. Here's an example: workers may use compressed air, which is a very expensive plant utility, to perform work that could be just as effectively performed by a brush or broom. In some instances, a less expensive utility such as flash steam could supplant the use of compressed air. An egregious example comes from one clever factory employee who "air conditioned" his workstation with streams of compressed air, which he enjoyed by simply tapping several nozzles into overhead air distribution lines.

- **Safety**. Lighting obviously contributes to the safety of working environments. We also use lighting to make a space more welcoming. We develop a habit of leaving lights on regardless of the space—storage rooms, break rooms, and worse, in rooms that are perfectly lit with natural daylight. As power becomes more expensive, we are forced to rethink these habits. Sensors and programmable controls are readily available to make the decisions that we humans can't (or won't) make.

The take-away is this: attempts to control energy use will almost always run afoul of someone's dependence on the service that energy provides. Managers attempting to reduce energy waste must recognize and overcome the need to use energy for unproductive reasons.

Okay, how do we improve our energy position? Let's recognize that top managers delegate activities, and energy management will be no different. Here's where it gets interesting. What if a staff person who is responsible for past energy performance is asked to investigate the potential for improvement? It's fair to question that person's objectivity. An effort to control energy costs now and in the future should first declare amnesty for the individuals linked to past energy performance (see page 69).

In moving forward, how can a lone energy manager control energy costs when consumption reflects the daily decisions made by operations, maintenance, engineering, and finance staff? This energy manager might get great ideas from workshops, conferences, trade press, and professional networks, but no one else from the facility is picking up the same messages. The energy manager easily becomes a maverick, swimming against the tide of a facility's disinterest (or worse).

This underscores the need for a team-oriented agenda and accountabilities. A major hurdle to overcome is the interdepartmental rivalries that are usually fueled by competition for budget dollars. Energy managers must somehow overcome the "silos" of departmental authority. For example, it's not uncommon for a procurement director to refuse to pay $10,000 for an energy audit that will identify many times that amount in potential savings.

The key is for the energy manager to demonstrate to other department managers "what's in it for them" should they cooperate with a facility-wide energy management effort. More specifically, this means demonstrating how energy efficiency's net benefits will filter down to a facility's bottom line. This means overcoming departmental resistance to spending a dime that in reality will result in the entire organization's saving a dollar.

The successful energy manager's agenda becomes a hub from which win-win solutions are distributed across facility departments. Technology is an inescapable part of this effort. So are human behaviors and procedures. Leadership coordinates technical with human agendas so that these create total value that exceeds the sum of the parts. Leadership is the underpinning of energy solutions.

Forward-thinking companies will change they way they use energy. They will often begin by rethinking their work habits and procedures. They will make an inventory of the gap between the efficiency of their current assets and the best available. A business plan will establish financial energy metrics, monitored at least once per month, to guide the content and pace of their energy business plan. They will document and replicate their success stories, and demonstrate their direct contribution to the bottom line of their business.

Top managers will quickly discover that energy use is as much a human issue as it is mechanical. To ignore the human component of energy cost control is to invite business risk. Awareness begets accountability. And with accountability, companies have the motivation to actively manage energy risk.

A checklist for advancing an energy management effort is offered on page 66, "The Seven Steps to Effective Energy Cost Control."

About Industrial Energy Policy and Programs

There is little doubt about the effect of rising energy prices on the profitability of American industrial activity. Energy consumers, producers, and government leaders are all compelled to do something about this challenge. As a result, energy issues have been a prominent feature of federal policy agendas in recent years. However, the policy-making process draws participants with highly varied perceptions of the causes of (and solutions for) today's energy market problems. Finance-minded business leaders usually anticipate a supply- or price-oriented solution. Engineers focus on advanced technology research and development. Government action emerges in the form of policies and programs, but whose agenda is reflected in these actions? Is it practical to expect lawmakers to solve industry's energy cost control issues? This section examines these questions. The conclusion is that policy and program initiatives are valuable, but the ultimate solution to industry's energy cost control challenges comes from leadership and accountabilities developed within industrial facilities themselves.

Let's be clear at the start of this discussion: legislative policy is a necessary yet insufficient tool for shielding the industrial sector from today's volatile energy markets. Industry—meaning the facilities that transform raw materials into the final goods that we consume—incurs energy bills that reflect a myriad of internal, day-to-day facility operating decisions. In general, manufacturers will assert that their internal business decisions are proprietary and off-limits to lawmakers. Accordingly, policy initiatives often focus on the supply side of the energy equation. Recent policies have attempted to ease restrictions on energy exploration and supply, which would ostensibly lead to lower energy prices. Supply initiatives, however, do nothing to address consumers' energy waste, which also inflates energy bills. As a practical matter, energy policy seeks to influence rather than control the decisions made by energy producers and consumers. These policies will, by their nature, have limited effectiveness in reducing industrial energy bills. Industry's relief from runaway energy expenses ultimately depends not on legislative action, but on business strategies developed and executed within facilities.

At the end of the first decade of the 21st century, energy issues are among the highest priority in many national policy agendas. The policy challenge is to coordinate a wide scope of energy themes, for example: the harvesting of fossil fuels; the overhaul and expansion of power generation, transmission, and distribution infrastructure; the commercialization of clean and renewable fuel technologies, and the adoption of efficiency technologies and practices that reduce energy waste. In the U.S., and after years of popular disinterest, national energy policy is evolving rapidly. National policy milestones include:

- the Energy Policy Act of 2005 (EPACT), which established incentives and programs for producing alternative energy supplies and technologies;

- the Energy Independence and Security Act of 2007 (EISA), which issued targets to advance alternative energy production as well as standards for boosting the overall performance of key energy technologies; and

- the American Recovery and Reinvestment Act of 2009 which, in its portions devoted to energy policy, essentially fortifies funding for the agendas established in EPACT and EISA.

Energy is a policy concern in most countries with mature or rapidly-maturing industrial sectors. Policymakers in these countries almost always struggle to reconcile the interests of maintaining an energy status quo versus the economic upheaval that may accompany the evolution to clean and efficient energy practices.

POLICY VS. PROGRAMS: WHAT'S THE DIFFERENCE?

Public policies are concepts around which laws, standards, and regulations are developed by legislative bodies (lawmakers) at federal, state, or local levels. Programs are administrative activities designed to implement policy. Public policies are developed from an original recommendation, or bill, that lawmakers then hammer into its final form through deliberation. The bill, in its final draft, must be ratified by an executive branch of government in order to become law. In the U.S., broad energy market policies are largely a federal concern. State and local energy policies tend to focus on building codes and construction standards.

In general, energy policies either restrict or encourage certain investment activity. Restrictions tend to focus on the terms and conditions for fossil fuel exploration, extraction, refining, and the interstate transmission of gas and electricity. Restrictions also impact the consumption of energy by prescribing performance standards for the design and operation of buildings and certain energy-using equipment. Policies often impose penalties against entities that don't meet prescribed criteria, which obviously require some kind of administrative enforcement function. Energy consumption is influenced loosely by policies that encourage investment, usually through tax incentives, in certain kinds of energy-related equipment. [2] Policies also authorize the development of energy-themed programs.

Programs are activities carried out primarily by government agencies. Each program reflects an agenda with clear themes, milestones, and objectives. Traditional energy program initiatives include technology research and development and market transformation activities (these will be explained below). Note that policies authorize programs, but program funding usually requires a separate legislative action. In other words, the policy act of authorization makes a "parking space" for a program concept, but the decision to allocate (or "park") funds is a separate issue.

ENERGY POLICY STAKEHOLDERS

Industrial energy consumption is a complicated matter that touches many decision-makers in a variety of ways, thus expanding the list of potential policy and program goals:

- Industry's corporate leaders are keenly aware of their rising energy expenses. These leaders demand relief primarily in the form of lower prices.

- Facility managers note the growing lack of skilled human resources needed to run their plants and keep pace with new technologies. They want training resources for existing staff as well as properly educated new employees.

- Vendors want to sustain industry's demand for the motors, pumps, insulation, controls, and other equipment that manufacturers rely on for their operations. Vendors want tax credits and other incentives to stimulate the market for their products.

- Facility engineers are responsible for the reliability of plant equipment. They evaluate the technology options for meeting production goals. Engineers want unbiased guidance to sort out the promises made by equipment vendors.

- Universities host much of the activity funded by energy research and development expenditures. They want sustained government support for new technology development.

- Gas and electric utilities must maintain the infrastructure that delivers energy to all consumers, including industry. Business planning is a difficult chore for utilities, since their customers' energy supply and demand projections must be sorted out before utilities can decide on their optimal level of infrastructure investment.

- Environmental advocates challenge the unnecessary depletion of natural resources, and seek to restrict energy-related practices that negatively impact air and water quality.

- Efficiency proponents remind us that energy depletion can at least be tempered through advanced technologies and best-practice procedures. Efficient use of traditional energy sources helps to buy time while advanced technologies and alternative fuel sources are being developed.

All of these stakeholders have a valid agenda. It's easier for lawmakers to craft individual policies for each of these agendas, but a policy framework that backs all of them simultaneously is problematic, as will be discussed below. None of these agendas, taken singularly, represents the comprehensive solution to industrial energy challenges. A comprehensive energy policy remains elusive, if only because of the varied needs and sensitivities of all stakeholders.

We can be confident that public policies will make more energy resources, technologies, and information available to industry. However, public policy can do very little to untangle the organizational disincentives and disconnects that prevent real energy solutions from being implemented. This suggests that the most actionable industrial energy policies are proprietary ones, promulgated by corporate entities as business plans for the benefit of their own facilities. A section beginning on page 100 describes ten companies that have done exactly that.

Energy management plans can be as many and as varied as the number of industrial facilities that dot the landscape. This is because each facility is uniquely characterized by its purpose, design, operations strategy, maintenance history, business objectives, and staff culture. As a consequence, there is no one-size-fits-all energy management protocol, nor is there a comprehensive policy design for facility-level energy management.

WHO SPEAKS FOR INDUSTRY? WHAT DO THEY WANT?

Lawmakers are responsive to the constituents and advocacy groups who bother to articulate their needs and wishes. If policy is the result of listening to constituents, then who speaks for industry? Is it the mid-level technocrats, or the corporate leaders of holding companies that own entire portfolios of manufacturing enterprises? This is a distinction of great consequence. Holding company directors generally know where their dollars go, but they may or may not fully understand the technical aspects of the operations under their control. They are happy to delegate technical issues—like energy—to mid-level managers. Lawmakers are more likely to network with corporate leaders than with factory technocrats. Corporate leaders, therefore, serve as industry's voice about needed legislation.

And how might corporate leaders express their energy wishes? Remember that as the U.S. produces ever-fewer engineering and technical degree holders, there is a growing disconnect between non-technical corporate leaders and the facilities they ultimately manage (see page 32). In other words, corporate priorities are increasingly set by people who have no idea how their facilities operate on a technical level. Many policy professionals are similarly uninformed. "Price" is one of the few concepts that is universally understood. Supply-oriented policy initiatives that seek lower energy prices will perfectly fit the expectations of finance-minded corporate leaders.

WHAT ARE INDUSTRIAL POLICY OPTIONS? HOW GOOD ARE THEY?

Initiatives to increase the supply of traditional fossil fuels and power generation dominated the U.S. energy policy agenda through the first half of the decade beginning in 2000. There are three main dimensions to supply-oriented energy policy:

- Regardless of how efficient a facility is, it will benefit from lower energy prices. For a given level of demand, more supply will ostensibly drive market prices lower. Opponents to this approach typically cite the negative environmental impacts that accompany the ever-more intensive extraction of fossil fuels. There are also compelling arguments about the dwindling supplies of fossil fuels and the need to develop alternatives. There is also concern over climate change that is attributable to fossil fuel combustion.

- Refineries and power generating plants are key components of energy infrastructure. In general, policies that reduce regulatory restrictions on the construction of energy infrastructure will presumably boost energy supplies and therefore reduce energy prices. Interests opposed to this approach are again typically concerned with environmental impacts. But even with regulatory approval, new energy supplies and infrastructure will take years to establish.

- Wind, solar, biomass and other renewable, non-fossil fuels are necessary components of the energy future. But one of the hurdles to ramping up these investments is the tangled mess that describes the current state of utility deregulation (see Appendix II). Before committing to these alternative fuel assets, investors need more certainty regarding the ongoing viability of fossil fuel markets, the future maintenance and overhaul of our national electricity transmission system, the state-by-state patchwork of utility distribution costs and regulations, and tax structures that directly impact all of the above.

Supply-oriented policy initiatives—focusing on traditional and alternative energy sources alike—will be only partial solutions to industry's energy cost woes. Other policy approaches may address industry's demand for and the efficiency of its energy consumption. A variety of demand-oriented policy concepts exists, and each is backed by one or more advocacy groups that work hard to make their agenda visible to lawmakers. However, each approach presents challenges when translated into policy:

- **Energy technology research and development (R&D).**

The development of advanced energy technologies is a task that few companies can pursue alone. Many technologies—like combustion, heat transfer, advanced materials, and controls—will have wide application across industries. Therefore, no one company or industry wishes to shoulder the burden of their development. The time, risk, and money that characterize R&D are best orchestrated through government-industry collaboration. Problems with technology R&D are it often takes years to come to fruition, the quality of labor and training available to the manufacturing sector doesn't necessarily keep pace with technology advances, and industrial facility managers are best advised to improve their current energy housekeeping before attempting to accommodate new technologies. The logic for this last point is simple: new capital investment projects are more likely to generate satisfactory financial results if they are complemented with energy-smart maintenance and operating procedures.

- **Greater efficiency**. Efficiency-oriented policies tend to use tax incentives and design standards to encourage (if not compel) specific energy-related investments. Equipment selections may include high-efficiency electric motors, pumps, or lighting. But in an industrial setting, sustained energy cost control is more dependent on whole-system designs, not isolated components. Facilities will achieve greater savings from an overarching energy-use strategy that harmonizes behavior and procedures to fully harness the benefit of efficient equipment. [3] Increased efficiency requires manufacturers to change the way they use energy and make energy-related decisions. Organizational complexity and inertia are huge barriers to making such changes.

- **Market transformation programs**. This concept emerged in the 1990s and is increasingly pursued in North America by states, provinces, and utilities. In brief, market transformation attempts to bring emerging technologies and behaviors into mainstream practice. This approach uses promotional strategies to effectively raise industry's demand for emerging technologies. It may encompass the other concepts described above, including greater efficiency, alternative fuel use, and investment incentives. Market transformation typically requires government-driven collaboration among energy end-users, energy utilities, and equipment vendors. The challenge is that market transformation programs influence but do not compel industrial decision-making. It is difficult for these programs to engage industrial organizations, since the appropriate facility contact is not an individual, but rather a team of decision-makers, all of whom have varying interest in energy issues and have other matters competing for their attention. Also, while vendors play a critical role in market transformation, care must be taken to not let them co-opt such programs for overtly commercial purposes.

Industry certainly needs energy price relief. But in addition to the "more supply" envisioned by many corporate leaders, industry needs technology R&D, efficiency, and alternative energy development if it is to achieve an effective solution to run-away energy costs. But note that lawmakers squeeze their work onto tightly-packed session agendas. Hammering a legislative bill into law requires trade-offs and compromise, so policy suggestions that are overly complex or vague have little chance for ratification. Simple, one-dimensional messages are easier to process in the policy arena. This works against the viability of a truly comprehensive energy policy.

REAL SOLUTIONS ARE INTERNAL SOLUTIONS

More supply, technology R&D, efficiency, and renewables—these are the energy policy options for lawmakers to ponder. Only so many federal dollars can be allocated to energy programs, per se, so these become competing options. Also, there are distinct advocacy groups that back each approach, and obviously these groups need money to operate. Each group in turn has its backers who will benefit if the government were to support its particular niche. This means that advocacy groups—each representing either more supply, R&D, efficiency, or renewables—are competing with each other. In addition, the need for simple advocacy messages discourages cooperation among advocates. It's very difficult to advocate what appear to be mixed messages. Note, for example, that many general observers tend to confuse "renewables" with "efficiency." This confusion becomes problematic when deciding where to allocate sponsorship dollars. Advocates are compelled to stick to their niche, because visibility lent to other agendas may be at the expense of one's own. Segmented energy policy concepts are valuable to individual advocacy groups, but are of limited value to industrial energy consumers. Unfortunately, a comprehensive energy policy, for which the whole has a greater value than the sum of the parts, has no backer. For policy advocates, it simply doesn't pay to take a comprehensive position.

Manufacturers can't expect policy alone to solve energy cost challenges. While lower fuel prices certainly help, the most effective way to control energy costs is to stop buying quantities that aren't needed. Each industrial facility must take control of its own energy fate through energy optimization plans that set goals, establish internal leadership, and assign accountability for results. Each facility is unique, and so is its optimal energy strategy. Public policy is weak medicine for energy issues, and certainly no replacement for good managerial decision-making.

Why are Corporate Energy Management Policies Needed?

Over the past two centuries, the size, pace, and complexity of industrial activity accelerated in parallel with the rise of corporate ownership and organization. To manage their operations and growth, corporations created entire departments of engineering, finance, marketing and other critical functions. Each of these departments now boasts a professional culture and influence well-defined by decades of academic training, professional accreditation, and practical application.

It is against this background, in the late 20th century, that industry began to encounter global competition for increasingly scarce resources. The energy management concept emerged as one consequence. Energy management, like any other organizational task, is subject to delegation. And obviously, energy management is still new, relative to other long-standing departmental responsibilities. In some organizations, the advent of energy management may create an unwelcome distraction for the agendas of other well-established departments.

Successful energy management therefore implies a need to accommodate change, not just of habits and technologies, but of the division of resources and influence within an organization. A truly ambitious corporate energy policy—one that actually seeks to reduce energy waste as opposed to merely paying lip-service to the concept—will instigate some measure of organizational change, to be facilitated by coordination among departments (see Chapter 4). That coordination, in turn, requires a sound policy that presents a vision, establishes goals, assigns accountabilities, and motivates action.

Corporate policy is especially useful for managing the change that is needed to overcome obsolete habits, procedures, and inertia. Effective corporate energy management is very much an exercise in change management—and not merely a collection of engineering "projects" (see page 85). A strong energy policy is not created for its own sake. Forward-thinking industrial corporations implement energy policies that support multiple objectives. These may include any of the following:

- To demonstrate administrative and legal compliance with respect to environmental liabilities, both current and future. Carbon is the big item here, but so are other direct emissions (fossil fuels used onsite) and indirect emissions (the emissions generated by the power plants that supply electricity). A good example of emerging regulations of this type is the mandatory greenhouse gas (GHG) reporting rule proposed by the U.S. Environmental Protection Agency in March 2009.[4] This rule would require facilities to keep records, starting January 2010, of the GHG it is responsible for producing. Penalties are proposed for non-compliance. This proposal is in response to the growing popular awareness of carbon emissions and their impact on climate change. The biggest and most controllable element of emissions compliance is energy use.

- To capture investment incentives that can improve energy performance and boost productivity. In the U.S., federal stimulus spending in 2009 provides the latest in an evolving set of incentives to invest in clean and efficient energy applications. While the disbursement of these funds is still a work in progress as of summer 2009, the limited funds are likely to reach the entities that are best prepared to receive them. To effectively capture these incentives, a company should (1) know its current energy profile, (2) know what the optimal energy profile SHOULD be, and (3) having a business plan for achieving that optimum. A corporate energy policy is that roadmap.

- To generate opportunities for new products and services. Wal-Mart is the corporate benchmark in this regard: as a corporate policy, they demand that their suppliers demonstrate clean, waste-reduced production processes, or else Wal-Mart doesn't buy from them. Procurement officers for defense department and government agencies are under executive order direction to improve the energy and sustainability performance of their facilities. It's only a matter of time before such criteria appear in the terms and conditions of work that they bid out to contractors. It's hard for a contractor to demonstrate compliance with these expectations WITHOUT a formal energy policy in place.

- To secure critical staff understanding and support for continuous energy improvement. Execution of any organizational effort hinges on its people. Energy cost control requires many staff to change the way they work and make decisions. In the absence of policy, there's very little to compel them to make those changes.

- To mobilize people and resources. A good policy means less work, not more. A good corporate energy policy sets the standard for taking action. It establishes clear roles, accountabilities, and investment criteria. A clear energy policy saves the company from wasting time and effort. Without such a policy, the company must deliberate energy improvements one at a time, mixed in with all the other core business decisions that have to be made. The energy agenda then becomes a stop-and-go process, and expect to "reinvent the wheel" many times over. Think of a corporate energy policy as a way to largely automate the decision-making process, removing the debate and speeding up the results.

Bottom line: a corporate energy policy has two major purposes: (1) to harmonize energy improvements internal to the organization, and (2) to prepare the organization for outside scrutiny and opportunities, such as regulatory compliance and market development. In other words, if energy policy is focused only on internal cost reduction, the glass is only "half full." A valuable energy policy is one that also helps to improve productivity and provides entry to new business opportunities.

Budgets Before Profits: Hidden Barriers to Energy Cost Control

In most industrial organizations, there is a woeful disconnect between daily decision-making and the profit motive. This is due in part to the size and complexity of the organizations themselves.

Because many skills and resources are needed to serve a production process, division of labor is a practical necessity. This is evident in the creation of departmental functions—and budgets. But in an environment of scarce resources, departmentalization can foster an internal, competitive dynamic that misallocates wealth.

Here's why: Budget development is as much about perception as it is the money itself. Budgets tend to be modeled on the previous year's actual experience. This puts the manager in a precarious position: under-spending this year could undermine the claim for next year's funding, while overspending may create the impression of waste or mismanagement. In effect, the department manager who economizes has just demonstrated the need for a smaller budget in the coming year. Managers tend to guard their budget dollars as a source of discretionary power. Over time, the annual cycles of budget development, defense, and execution yield a culture of hoarding.

Within the typical industrial organization, certain barriers to energy cost control are a consequence of departmental competition for budget dollars. Energy control activities and costs may be delegated to a "facilities" department, or wherever engineering and maintenance tasks are handled. An industrial facility manager ensures that buildings, manufacturing processes, and attending staff have the heat, power, ventilation, and other services needed to function effectively. These activities are often perceived as secondary in importance, relative to the core business of manufacturing products and meeting production goals. Accordingly, facility managers may be at a disadvantage when competing for internal budgetary and analytical resources. "Success" for a facilities manager means keeping emergency failures to a minimum. By definition, emergency issues are unpredictable in size and frequency. Given the choice between emergency preparedness and the efficiency of ongoing operations, many facility managers are hesitant to spend money on "fixing things that aren't broken." This allows energy waste to persist.

Everyone else carries on business as usual without regard to the energy expense implications of their actions. The facilities manager alone would be responsible for reversing the wasteful choices of others. This could be the job of a proverbial Sisyphus, never-ending and without reward. Unless everyone is accountable for energy use, an energy manager's effectiveness is severely limited. Under these circumstances, energy waste will prevail, directly reducing the financial return available to shareholders.

LOOKING FORWARD THROUGH THE REAR-VIEW MIRROR

We're all familiar with the monthly budget review meeting. This is when the general manager sits down with department heads to compare the latest month's financial results to the organization's operating budget. This is one way in which department managers are held accountable for their year-to-date fiscal performance. It's a very common and well-intentioned business habit. It's also potentially one of the most damaging, because the actual-to-budget review process focuses on the past at the expense of the future. It's like trying to steer a car by looking in the rear-view mirror. And as you will see, it is a big reason why organizations perpetually fail to take meaningful control of their energy costs.

While the organization as a whole attempts to make money, department directors are primarily concerned with spending money for materials, labor, utilities, support services, and the like. The monthly budget review is a discussion of variances—particularly those instances where spending is on a pace to exhaust funds before the end of the fiscal year. Top management rewards people whose annual expenses come in under budget. However, a preoccupation with this year's budget may be at the expense of potential savings that can accrue for years to come.

Here's a realistic example of how the dollars get in the way of energy solutions: A facility has an annual energy expenditure of $10 million. It forfeits the opportunity to identify $1-2 million in potential savings, simply because someone wanted to avoid spending $10,000 on an energy audit. Why? Because that person doesn't want to miss the opportunity to pick up a $500 bonus for coming in under budget for the year.

We allow this to happen because this year's budget is largely derived from last year's actual fiscal performance. History provides some useful insight, but it can also obscure future potential.

Here's another example that illustrates the "rear-view mirror" management approach to moving forward. Let's say a facility has yet to adopt energy-efficient technologies, behaviors, and procedures. This means that it habitually buys more energy than is actually needed, because waste is built into its operations. The budget account for energy, then, is inflated to accommodate these inefficiencies. As we saw on page 10, energy losses add up to about 40 percent of the total energy delivered to U.S. manufacturing facilities as a whole. Stated differently, the typical manufacturing facility must inflate its energy procurement budget by an additional two-thirds to account for the energy that it will eventually waste.

Consequence: The monthly budget review routine forces people to manage this year's cash flows based on last year's expense determinants.

Possible solution: Break down annual energy expenses into two separate line items. One represents the value of energy that will be actually applied to perform useful work. The second line item represents energy that will be wasted. How do you allocate energy expenditures into these categories? The answer is found by conducting an energy audit that thoroughly evaluates energy inputs, uses, losses, and potential consumption improvements (see page 71). While industry averages are generally helpful, the most reliable indication of any single facility's energy flow depends on a proper energy audit—the more thorough its scope, the better.

Without distinguishing between energy applied and energy wasted, department directors so often conclude that they "don't have the money for energy improvements." In fact, they DO have the money, and it's earmarked for future purchases of energy that will be wasted. The account for energy waste (a budget artifact directly related to past performance) is in reality an account from which energy improvements should be budgeted. The energy waste line item also brings attention and urgency to the issue at each and every monthly budget review. Managers can use the "energy waste" account to either make energy savings or buy energy that ends up being wasted (see page 121). Dollars from the "energy waste" account are devoted to energy improvement projects only when the cost to save a unit of energy is less than its purchase price per unit. This is one line item that actually forces managers to look forward, and not in the rear-view mirror, when planning energy consumption.

Is Management to Blame?

Don't worry, this is not an indictment. Top managers have a company to run, and they can and do make money despite inefficiencies. However, the burden of energy waste, lost income, and increased exposure to operating risk are increasingly hard to bear in a globally competitive economy. Energy has become a subject worthy of strategic management, a fact that has yet to be fully appreciated.

Research conducted by the Alliance to Save Energy [5] examined the organizational aspects of industrial energy management. From this effort, certain barriers to energy efficiency were identified:

- **Executive disconnect**. Organizations don't change without concerted direction from the top. The adoption of energy efficiency exemplifies that sort of change. But to start this process, energy has to be on the executive radar screen. Usually, it's not, because executives are fully tasked with revenue/profit targets, product development, and legal or regulatory compliance issues. Energy use is rarely noticed or questioned above middle management levels. When executives are faced with energy challenges, they quickly delegate it to technical subordinates who are usually poorly prepared to tackle the organizational changes that real energy solutions require.

- **Inconsistent leadership**. The rotation of management within companies often prevents the hard decisions from being made. "Not on my watch" is often the response to improvement proposals that won't pay off until after the current general manager's tenure is over. Energy improvement efforts can also lose momentum simply because key personnel are reassigned or because management changes the organization's priorities.

- **Lack of common understanding**. The concept of energy efficiency must be mutually understood by industry's business directors and their subordinates across a number of facility departments. Without mutual understanding, the direction to become energy efficient probably will be distorted and diluted throughout an organization, even if it is a corporate initiative. A common understanding will clarify the expected outcomes and solidify support for the activities that bring results. Organizations that pursue energy efficiency without a clear understanding of what it entails will assign responsibility to the wrong people, or not empower enough of the right people, and lack appropriate coordination among staff.

- **Lack of staff and management awareness**. Plant personnel don't always make the connection between energy choices and money. For example, compressed air leaks are often overlooked because "air is free," although this conclusion ignores that fact that five to seven horsepower of electricity are consumed to generate one horsepower of compressed air. Steam system management is susceptible to similar thinking. Plant operators who assume that scrap rates are of no importance "because scrap can be melted down and used again" are not considering the excess energy consumption that this practice requires.

- **Lack of procedural coordination**. In most companies, an under-informed pursuit of energy efficiency is bound to conflict with a wide variety of existing operating procedures. For example:

    — **Restrictive procurement requirements**. While energy-smart accounting usually recognizes total cost of ownership, procurement staff may be motivated to focus only on lowest up-front costs.

    — **Outdated accounting procedures**. When a single utility meter measures total energy consumption for an entire facility, then accountants must arbitrarily assign costs across departments. This means prorating the utility bill to each department per some proxy, such as number of square feet or by number of personnel. This accounting scheme will disguise the sources of energy waste as well as the impact of any specific improvements. Improper allocation of energy costs may distort financial decisions such as product pricing, income and tax declarations, production mix, compensation and bonuses, and capital investment allocations. Today's advanced energy metering technologies can monitor actual consumption by substations within a facility, improving department managers' abilities to control their energy costs.

- **Absence of energy-related accountabilities**. An industrial company's organization chart features departments that are designed to support core business objectives. Energy efficiency is not typically integrated into individual employees' job descriptions. Even if staff agree in principle to save energy, the required effort can still be derailed by existing routines, habits, and procedures. Accountabilities must be modified to recognize waste management principles in general, which include energy efficiency. Without these accountabilities, the company will find it difficult to attain any lasting efficiencies, even if ordered to do so by a chief executive officer.

- **Fear**. Does admitting the need for efficiency improvements become evidence of ineffective job performance? Many plant personnel think so. People are also afraid that energy management means more work to distract them from their already loaded schedules.

- **Denial (or *don't ask, don't tell*).** It is easy for top managers to be lulled into complacency about energy and other support functions with which they are not familiar. Management indifference effectively abdicates control to trusted subordinates who know that it is better to report good news than bad. Who is a 35-year-old general manager to question the report of a power plant superintendent with 20 years on the job? These territorial relationships are barriers to energy efficiency, especially when tenured staff explain that "this is the way we've always done it."

- **Lack of resources**. Because of limited time, money and skills, and with management accountability sometimes tied to short-term results, deferred maintenance is the order of the day. To save money, some companies will release well-compensated, skilled workers, especially from non-core activities like energy support. The remaining staff are ill-prepared to seek, promote and maintain energy system improvements.

The most durable barrier may simply be an organization's business culture. Few corporate leaders, if any, "save" their way to the top. Their bias is for short-term revenue making, not cost saving. This thinking is evident in capital budgeting decisions, where growth-oriented projects are favored over expense-reduction initiatives. Decision-makers who dismiss energy efficiency overlook opportunities to grow revenue through the redirection of energy waste to more productive purposes.

It bears repeating that there are no villains seeking to create industrial energy waste. There are only organizational procedures, work habits, and accountabilities that have stood still for years while the relative costs of time and energy have changed.

How to "Do Nothing" About Energy Costs

Is "doing nothing" a valid way of dealing with energy costs? Well, it's a particularly attractive choice if you are determined to operate the plant the way you've always done it. You don't have to line up in the capital budget process to get money for energy improvement projects, and you don't have to force people to change behaviors and procedures. You avoid the risk of getting fired because the project that you fought to get approved failed to meet its projected performance measures. So instead of taking on the sharp pain that accompanies change, you settle for the dull ache of income lost to energy waste.

Doing nothing saves you from many problems, but it does not shield you from the fact that financial statements don't lie. Top managers may not track energy, but they do track money. So as it turns out, "doing nothing" about energy eventually puts you at risk. Here are some suggestions for how to cover your tracks:

Blame your energy expenditures on rising energy prices. You don't set the price for fuel and power, the market does. You're at the mercy of the market. Front-office people don't always understand how your mechanical systems use energy, but they certainly understand price. To strengthen your case, start charting energy prices over time. Pictures like this are worth a thousand words. Use this price data to explain why your energy expenses keep going up. You want to disguise for as long as possible the fact that energy bills are a function of volume consumed as well as price.

Remind your management that your competitors have to buy energy, too, so they have the same problem and therefore the playing field stays level. Don't mention the fact that your competitors may be addressing their energy waste, because that would shoot a hole in your "level playing field" claim.

Cut other expenses. As your energy bills go up, the money to pay them has to come from somewhere. The favorite targets are maintenance (which can always be deferred, right?) and labor, assuming you haven't already trimmed payrolls to the bone and then some.

Pass the costs increases on to your customers. There's only one problem with this: competition. Customers can take their business elsewhere if they don't like the price you charge. Compensate for product price increases with enhanced service that presumably doesn't add to your expenses in other ways.

Alter the mix and quality of subcomponents that make up your final product. For example, some creative manufacturers are settling for more brittle grades of plastic or metal hardware that are not heat treated or thoroughly plated for corrosion resistance. That's one way to hold down costs, but you run the risk of alienating customers with junk products.

Change your accounting of energy costs. In other words, change the way facility-wide energy bills are broken down for assignment to individual departments. Many organizations have one meter for electricity, for example. An accountant merely prorates the total bill over the departments by some artificial measure, such as number of employees or square feet—measures that usually are poor indicators of energy use, but they are convenient for the accountant's task. Use a spreadsheet to model energy cost allocations under different scenarios. Find an allocation that works to your benefit (that is, hides your energy waste), and persuade the finance controller to adopt your approach.

Get political and start lobbying your legislators to ease restrictions on energy supply. Presumably, if we dig, drill, refine, and generate more power, the increase in supply will drive prices down. Corporate officers and politicians understand "price." Any discussion about how energy is used (and wasted) gets too complicated. By focusing on price, you keep it simple, and therefore increase your chances of getting political support. But before going down this route, you should see page 42, "About Industrial Energy Policy and Programs."

The Seven Deadly Sins of Energy Cost Control

7. Believing that the financial impacts of energy decisions are reflected only in utility bills, ignoring the wider impact on raw material consumption, labor and space utilization, process set-up and cycle times, product quality, and competitive positioning.

6. Believing that energy prices are the sole determinant of energy expenses, while confusing "efficiency" with "environmentalism."

5. Implementing one energy "project," declaring victory, and moving on to other matters because "the energy problem is solved."

4. Relying on technology and equipment alone to improve energy performance, without recognizing the role of behavioral and procedural change.

3. Delegating the energy challenge to a single person or department before mapping out the causes of energy waste and the potential solutions.

2. Failure to establish clear lines of accountability so that staff are responsible for the energy implications of their actions. Failure to distribute the costs and benefits of plant-wide energy improvements across departments.

1. Fear of admitting that energy waste exists, yet holding individuals responsible for waste that is really attributable to management system failures (see all points above).

The Seven Steps to Successful Energy Cost Control

1. Top management must demonstrate its clear and durable intent to progressively improve the energy performance of the entire organization. Management should also declare amnesty and hold individuals blameless for past choices that caused energy waste.

2. Develop energy-use profiles at the facility level. Benchmark and regularly track the volume of energy that the facility *should* use. Devise simple, clear metrics for this purpose.

3. Identify and remove organizational disincentives to changing the habits and procedures that lead to energy waste. Ensure that the costs of energy use, and the benefits of increased efficiency, are distributed across all departments.

4. Establish leadership, accountabilities, and a protocol for effectively remediating lapses from optimal energy use.

5. Regularly document and communicate energy's contribution to business performance. Use business language instead of technical language to accomplish this.

6. Seek, encourage, and reward ongoing innovation for harvesting wealth from energy use.

7. Make energy-smart criteria integral to every-day operating, engineering, and procurement decisions. If energy cost control is still perceived as a distraction, then the organization has yet to create the incentives to truly manage energy risk.

Endnotes

1. United States. Department of Energy Industrial Technologies Program. *Energy Use and Loss Footprints.* http://www1.eere.energy.gov/industry/program_areas/footprints.html. Accessed March 2, 2009.
2. Chapter 12 explains why manufacturers respond slowly and partially, if at all, to incentives to adopt new and efficient equipment.
3. http://www.ase.org/content/article/detail/3177. Accessed March 2, 2009.
4. http://www.epa.gov/climatechange/emissions/ghgrulemaking.html. Last accessed August 1, 2009.
5. http://www.ase.org/content/article/detail/3174. Accessed March 2, 2009.

Chapter 4

Change

"I'm all for progress.
It's change I object to."
—Mark Twain
(1835-1910)
American humanist, humorist, satirist, author, lecturer, and confidant to statesmen and industrialists.

Twain was masterful at divining incisive wit from colloquial dialogue, thus imparting philosophy that resonates easily and effectively with audiences to this day.

Wanted: Amnesty for Yesterday's Energy Waste

Manufacturing leaders need to declare amnesty for plant personnel who are on the front lines of energy use. Energy is not wasted on purpose or by design—instead, it is the natural consequence of having too few people monitoring too many dynamic business forces. As many facility-level staff will say, "It's all we can do to keep the 'car' on the road, much less tinker with its efficiency."

Individuals should not be reprimanded for outcomes that are really caused by organizational gaps in energy decision-making. As page 38 shows, the causes of energy waste are deep-rooted. These causes took time to manifest, and are never limited to one individual's actions.

People are afraid that they will be held culpable for longstanding habits and procedures that are not of their making.

Workers at the lowest levels may understand better than anyone else the turf issues and contradicting priorities that prohibit changing the way energy is used and controlled.

For example, it is not uncommon for one department to undertake and absorb the expense of an energy-saving initiative, only to have the savings reflected on another department's books. To add insult to injury, the first department gets chided by management for incurring unusual expenses. Organizational disconnects like this allow industrial energy waste go unchecked.

Energy solutions will depend on cross-departmental communications that may be unprecedented in many facilities. A procurement director, for example, may be required to select products offered by the lowest of three bidders. An energy-smart procurement policy would prioritize equipment options that impose the lowest total cost of ownership, which may be different from option with the lowest up-front cost. Will the procurement director's performance criteria be amended to accommodate this? There are potentially many organizational barriers like this.

This explains the call for amnesty within industrial facilities. A manufacturing facility that wants to reduce energy waste must take a team approach to the process—identifying barriers and the potential means for circumventing them. This early stage of the energy planning process should make ensuing technical and procedural discussions that much easier to conduct.

Industry cannot afford energy waste, nor can it afford to waste time by trying to avoid or assign blame for yesterday's decisions.

An Entitlement Under Fire?

A provocative story, entitled "Get Healthy or Else," was on the February 26, 2007, cover of *Business Week*. It's about employers that find it increasingly expensive to provide health insurance coverage in the face of double-digit annual cost increases. One company, Scotts Miracle-Gro of Ohio, is responding by pressuring, counseling, and coaching its employees to lose weight and quit smoking. Note that this is a corporate effort initiated at the very top of the company by the CEO. As you can imagine, there is a subtext to this story having to do with individual rights. There are legal challenges, so it's not yet clear that the Scotts example can be sustained for long.

Scotts, you see, is tired of throwing money at a problem that is already out of control. The alternate strategy is to focus on prevention. What a concept!

You would think that companies are more firmly within their rights to crack down on energy costs. After all, by attacking energy waste, a company acts on behalf of its own assets, resources, and procedures. Be it health care or energy, the dollars at risk are not inconsequential. When you address habits (and that's the common ground between these two subjects), corporate leadership may make all the difference. If you read the *Business Week* article, you'll see how top leaders walk the plant floor to personally reinforce policy—face-to-face, one employee at a time. Tackling energy waste may require this kind of leadership, too.

Questions and Answers About Energy Audits

You can't manage what you don't measure.

The process of energy cost control starts with an energy audit. [1] An energy audit identifies energy consumption that is in excess of what is needed to adequately serve a facility. A report of findings should list potential improvements, showing the volume of anticipated energy savings and an estimate of the cost to achieve those savings. The audit process itself doesn't fix anything. It merely provides a "roadmap" that shows where savings are. At the very least, an energy audit itemizes the costs and savings attributable to each improvement recommendation. [2]

WHY WOULD ANY BUSINESS WANT AN ENERGY AUDIT?

If the purpose of the industrial organization is to make money, it should manage both revenues and expenses. Energy audits are the blueprints for recapturing the wealth that is embodied in energy use. A useful energy audit will present:

- An energy balance that compares energy inputs and outputs for the entire facility
- An analysis of current tariffs and purchase agreements related to energy acquisition
- Ways to optimize the operations and maintenance of current assets and equipment
- New technologies that may be adopted
- Potential for alternative sources of fuel and power
- Measures of cost-benefit for specific improvements
- An explanation of available incentives, rebates, training resources, etc.
- A summary of environmental liabilities and rewards related to current and potential energy use (see page 157)

WHAT KINDS OF PROFESSIONALS PERFORM ENERGY AUDITS, AND WHAT IS THEIR FOCUS?

The auditor should have a technical understanding of heating, cooling, ventilation, lighting, combustion, motor drives and other technologies. The energy auditor typically recommends a list of hardware upgrades or changes, with each discrete opportunity becoming a potential project. Traditional energy audits are performed by engineers for engineers, so the reports tend to be written in a technical language.

IS THERE A PRACTICAL DIFFERENCE AMONG ENERGY AUDITORS AND THE RESULTS THEY PROVIDE?

Put it this way: if you get ten different audits, you will get ten different results. There will probably be significant overlap, but each report reflects the experience, skills, and biases of the professional who prepares it. There are certain audit standards to which state-licensed professional engineers (P.E.s) are held, but even then there's room for individual interpretation. Energy audits are not a commodity, and you get what you pay for.

WHO PROVIDES FREE ENERGY AUDITS, AND WHAT SHOULD YOU EXPECT FROM THESE?

There is no free lunch. A "free" audit is often a front for a commercial agenda. If so, the provider of a free audit is not truly working for you and your benefit.

- Some utility companies provide free audits, usually because regulators order them to do so. These utility programs, however well-intended, are chronically under funded. Do not expect a comprehensive analysis.

- University or government agency audits. You can get some value here, but again, these providers are under funded. They don't have the resources to perform extensive study and analysis. At least there is a lack of commercial bias, so it's better than nothing.

- Equipment vendors offer free audits. If you really expect someone who makes money by selling his equipment to give you an objective, unbiased audit, then go for it. Having said that, you may learn something if you compare the results of two or more free audits as offered by competing vendors.

Here's a question for *you*: Do you not have the money to pay for an energy audit, or do you not *have the authority to spend the money* for it? Unfortunately, energy audits will recommend things that cost money or require changes to operating procedures or staff behavior. Will you have the authority to pursue those courses of action? An energy audit will identify opportunities to shape your utility bills for the next five, 10, 20 or more years. The value of the energy wasted over that time will dwarf the amount you could have spent for a proper energy audit. Do you want energy improvement decisions to be based on the results of a free audit, or the results of a proper analysis?

IS THERE A SHORT, QUICK, CHEAP WAY TO CONDUCT AN ENERGY AUDIT?

Sorry. No one can stick his head in the door and guess in advance how much savings are in a facility. It takes time to prepare for and conduct an energy audit, then to analyze the resulting data. It involves a physical walk-through of the facility as well as the study of blueprints, mechanical drawings, and production or operating schedules. The analysis needs to be performed by qualified individuals with the appropriate skill set. To get the most from an energy audit, your staff should be aware of this process and prepare accordingly to accommodate the auditor's study.

WHAT'S THE RETURN ON INVESTMENT FROM AN ENERGY AUDIT?

Zero. Zip. Expecting an energy audit to save you money is like sitting on a road map and expecting it to take you from New York to St. Louis. The road map itself does not do the work—you do. The road map helps you prepare so that you can find the most effective route that economizes your time and money. Energy audits do the same. While the return on investment for an energy audit is zero, it's the recommendations themselves that provide returns, but only if you decide to implement them.

IS AN ENERGY AUDIT CRUCIAL TO BUSINESS COMPETITIVENESS?

It depends: you don't need it if you can always pass all your costs on to your customers, can always boost your sales volume to make up for smaller profit margins, and are immune to regulatory compliance and scrutiny from shareholders and employees.

SOME PEOPLE IN THIS ORGANIZATION RESIST HAVING AN ENERGY AUDIT. WHAT SHOULD WE DO?

An energy audit should be approached as a treasure hunt, not a witch hunt. The challenge in conducting an industrial energy audit is to manage the expectations and interests of many stakeholders. It's not unusual for facility staff to be threatened by an outsider's investigation. To be successful, communicate in advance to all stakeholders, clarifying the intention and outcomes sought from the energy audit experience. Be prepared to explain to staff "what's in it for them" should the audit be conducted. The audit report should recognize what the facility does right as well as areas for improvement. A presentation of findings should clearly describe the impact of proposed energy solutions on business performance. Top managers don't need to be involved in the technical aspects, but they should be aware of the business potential of the recommendations. Their leadership will help secure the cooperation needed to conduct the audit.

OUR ORGANIZATION SIMPLY WON'T APPROVE AN ENERGY AUDIT, BUT WE STILL WANT TO REDUCE ENERGY COSTS. WHAT CAN WE DO?

If all else fails, simply scour the Internet for energy saving tips. Make a list of potential improvements, and prioritize these the way you think best: do what's cheapest, do what's easiest to accomplish, etc. Be sure you benchmark your energy consumption over time so that you can measure the impact of your improvements. How much potential savings will you capture this way? And how long will it take? To answer those questions, you will need to secure... an energy audit. Your last resort is to "do nothing" about your energy costs. It's a valid approach, but it has its costs, too. For that topic, see page 63.

SO I HAVE AN ENERGY AUDIT. NOW WHAT DO I DO?

A good first step is to translate the highlights of the audit report—a technical document—into non-technical terms that will resonate with non-technical business leaders. The report's simple list of recommendations should be evolved into a business plan for action. The audit report says nothing about which department in your organization should pay for improvements, which department will book the savings, or if your staff or an outside contractor is best able to do the work. The report says nothing about the disruption that such work might cause. How will you prioritize projects? How will you measure the financial impacts of each? These are the questions that an energy business plan should answer.

So here's the take-away message about an energy audit: Don't look at it only as a cost. If you do, you can certainly save money by not securing the audit in the first place. Rejecting the audit opportunity is a very high risk strategy, because each dollar saved by avoiding an energy audit can cost many more dollars in energy waste. Understand that an audit is only the first step. Energy costs are controlled by actually pursuing the opportunities that an energy audit recommends.

Move Forward or Fall Behind

When it comes to energy management, there are only two choices: move forward, or fall behind. There is no "idle" setting. Failing to take action has its costs, too. This is the implicit choice of companies that choose to continually resist energy cost control.

If you choose to "do nothing," it costs nothing, right? Let's think about it: you may have already obtained an energy audit, but that by itself accomplishes nothing. All savings from an energy audit remain forfeited until you actually implement the recommended improvements. Many good proposals languish while the concept must be sold to a variety of decision-makers within the organization who may or may not understand the proposal's technical aspects. This internal "sales" process can take months, because it has to find a place on a series of meeting calendars. The longer it takes to ponder energy improvements, the greater is the chance that capital will be diverted to some emergency repair item that inevitably arises. No wonder the "do nothing" approach is so enduring.

Here's how you can fall behind by simply standing still: the nominal efficiency of energy-consuming machinery inevitably erodes over time as heat and friction take their toll. Maintenance expertise similarly erodes with staff attrition and turn-over. If best-practice procedures are not documented, people scramble to reinvent knowledge that should have been part of standard operating procedure to begin with. In the meantime, new best practices and technologies are appearing all the time. Is anyone scouting for these opportunities? Does it still make sense to do things "the way we've always done them?"

The good news is that energy management does not have to happen all at once. Successful programs start out in first gear, seeking early victories to give the organization confidence in what they can do. Momentum pushes their efforts into second and third gears, as decision-makers become comfortable not just with hardware changes, but with team-based decision-making. They've reached full speed when energy-saving initiatives become part of standard operating procedure. Then—and only then—does the organization occupy the driver's seat when it comes to moving forward with energy cost control.

Energy Management: Two Philosophies, Two Outcomes

Show me a business factor that is low-cost, stable, predictable, and non-controversial, and I'll show you a factor that is easy to manage. Until recently, fuel, power, and other utilities were a great example of easily-managed inputs. But as we all know, everything about energy is changing: its availability, its cash flow impacts, and the legal and environmental consequences of its use. Energy use is no longer the one-dimensional factor of years past. As energy becomes more valuable, fuels and power become synonymous with wealth.

Against today's energy landscape, two distinct management philosophies characterize the way energy is managed by industrial or large commercial organizations. For discussion's sake, let's refer to these philosophies as "business as usual" versus "forward thinking." The two philosophies have powerful implications for business performance.

The business-as-usual philosophy was developed during an era when energy was still cheap and simple to use. Historically, energy consumption was an internal facilities issue. Energy procurement choices had little, if any, linkage to consumption decisions. A facilities agenda could be easily isolated from that of the larger organization. The facility management goal was quite simple: achieve 100 percent availability and reliability of heating, cooling, and other energy services. Cheap, reliable energy services at the flick of a switch were the mark of excellence in facility management. For the most part, facility managers went unrecognized until some failure in energy systems became evident. This philosophy obviously breeds a conservative mind-set, one that resists change to existing technologies, assets, and procedural routines. The business-as-usual facility manager becomes proactive only as needed to maintain a functional status quo. For a variety of reasons, the business-as-usual approach is alive and well in countless organizations, despite a growing need for flexibility in the face of volatile energy markets.

The forward-thinking organization's approach to energy recognizes and responds to change. Energy—impacting the organization's triple bottom line of economics, social responsibility, and environmental impacts—now has consequences for revenue, expense, and risk performance. In the forward-thinking organization, energy decision-making involves staff in operations, maintenance, engineering, finance, procurement, environmental/health/safety, and marketing and product development. Not that these dissimilar professionals are active participants in the boiler room; rather, energy-smart criteria are folded into their standard operating procedures. This organization monitors energy use as well as the evolution of related technologies, regulations, and cash flow opportunities. Rather than clinging to a functional status quo, the forward-thinking organization effectively "connects the dots" between energy and its business performance. In other words, staff are empowered to make choices that use energy to their organization's best advantage. Goals, accountabilities, and top-level leadership support this vision.

Today's energy markets impose unprecedented challenges—and opportunities. Business organizations can and do evolve in response to this environment. Show me an organization that clings to the business-as-usual approach to utility management, and I'll show you an organization that loses opportunities to conserve, invest, and ultimately grow its wealth.

Some Can, Some Can't (...control energy costs)

Some facilities can do it. But it's not easy. Controlling energy costs means making changes, and change is a tall order for the staff of many facilities. You can see why, especially when rosters are trimmed to the bone and people accumulate multiple responsibilities. Over-stretched workers crave routine—it is relief from the "fire drills" that often characterize today's lean-and-mean production environments. Nevertheless, some organizations will embrace change because its people realize that their livelihood depends on it.

One example comes from a plant in central Pennsylvania in early 2007. A positive, can-do attitude was pervasive during a walk-through. Staff people greeted visitors with smiles. The facility was clean and well-lit. The common areas and rest rooms were obviously well cared-for. Glass cases in the lobby proudly displayed their products as well as trophies from the little league teams that the plant sponsors. In short, there is evidence of a quality covenant between management and the facility staff—a relationship of mutual respect and concern for their facility and its current and future viability.

There is certainly plenty of work to do there to get an energy management discipline in place. But at least these early observations suggest that this facility has many of the intangibles needed to be successful.

Other facilities can't do it.

The changes required by energy management are too distracting for some people. Why? Because they do not think of energy as wealth. Without performance metrics, there's no way to link energy use to individual decisions. Without accountability, there is no cost control.

One example comes from a precision-tool manufacturer in Maryland. Its long-time facility manager had an energy strategy that centered exclusively on shopping for electricity at the lowest price, a privilege available in Maryland's deregulated utility market. For this, he gladly enlisted the help of an independent advisor.

The advisor had always encouraged this facility manager to pursue a business plan for energy improvement. Beginning with an energy audit, this plan would inventory the gap between current energy use and the optimal consumption that would be made possible by changes in technologies, behaviors and procedures. The facility manager refused an energy audit unless it was provided for free.

In mid-2007, the trade press announced that this facility was purchased by a holding company. The advisor contacted the facility manager again, pointing out that new ownership signaled a time for staff to be on their toes. After all, new owners usually bring with them an agenda for change. The energy business plan concept was suggested again to the facility manager as a way to demonstrate to the new owners his vision and preparedness for his accountabilities. And again, he refused the idea. Energy audits and planning were apparently not worth paying for.

The latest news about the facility manager surfaced as gossip at the energy advisor's 2007 holiday party. This manager had been laid off, and was calling around about new employment possibilities. Was there direct causality between his lay-off and his refusal to get serious about energy management? We don't know. Hopefully, he will be employed soon. Maybe next time, his approach will be somewhat different.

His livelihood may depend on it.

Endnotes

1. The term "audit" may cause some discomfort to business managers who are accustomed to the pecuniary implications of tax or safety audits. Realizing this, there are some industrial energy efficiency proponents who prefer the term "energy assessment" or "energy profile." A decade into this new century, we have not yet reached the tipping point in this evolution of semantics. If an analysis of Internet search terms that bring visitors to the author's blog is any indication, "energy audit" is the preferred term in the minds of most people who are investigating the subject.
2. Thanks to Don Wulfinghoff of WESINC, whose teaching is reflected in this chapter.

Chapter 5

Developing an Energy Strategy

"In preparing for battle,
I have always found that plans are useless,
but planning is indispensable."
—Gen. Dwight D. Eisenhower
(1890-1969)
34th President of the United States
Commander, Supreme Headquarters
Allied Expeditionary Force

Despite never seeing combat, Eisenhower was revered as a war hero, in part for his skillful and diplomatic leadership of multi-national armies.

Do the Right Thing!

Virtually all industrial organizations are aware of increasing energy costs, but few have found a consistent solution. One explanation: we are so focused on "doing things right" that we **fail to do the right things**.[1]

The traditional approach is to "do things right." It goes something like this:

- Like many companies, we consider ourselves "lean." For us, this simply means people wear many hats. Fire drills can dominate our days, so we seek relief by establishing as much routine as possible.

- We foster internal competition for resources, especially at budget time. Like it or not, we make decisions to optimize departmental performance, not business-wide performance. We avoid spending departmental dimes that would actually save dollars for the whole company.

- To survive professionally in a lean environment, don't make waves. Stay "inside the box" that (we hope) gives us some routine. Need to solve a problem? Do it on your departmental turf with your own resources.

- Like everything else, energy is an issue to be delegated—usually to a technical person or department. Not only do we expect technical people to develop energy solutions, we expect them to do it inside THEIR box. This means implementing isolated projects while everyone else carries on business-as-usual.

- When delegating, we get what we pay for. The lower we delegate a task in the organization, the more localized and temporary the solutions become.

Result: our desire for routine and to not upset the status quo means that we never solve the root causes of energy waste—waste that becomes more costly as energy prices escalate.

What will allow us to effectively control energy? Or in other words, how can we "**do the right things?**"

- First, understand that technical projects alone cannot solve problems rooted in management systems and culture. Management procedures and systems, as well as staff's behavioral habits also determine energy performance.

- Next, re-read and accept the first point. Take heart, because technology is still part of the solution.

- Delegation should not be the first step in solving energy problems. Instead, map your energy decision-process to identify the choices, roles, and performance metrics that link energy inputs with performance targets. Establish goals based on identified savings potential. Tune up the system. Then develop accountabilities to keep performance on track. Scout for new technologies and practices that allow the bar to be moved even higher. Welcome to *continuous improvement*.

First: learn to do the right *thing*... then do it *right*.

An Engineering Project or a Management Process?

What does it take to control energy costs? Is it a capital investment project? Or is it a management process? Before answering that question, think about your retirement plan. Do you randomly invest large chunks of money exclusively in high-risk equities, or do you diversify assets and mitigate risk through dollar-cost averaging? If you understand the logic of investment portfolios, you understand what an organization-wide energy management program can accomplish.

The project approach comes naturally to engineering-minded decision makers. This comment is not meant to detract from engineers—in fact, their skills remain central to the sustained reliability and efficiency of any production facility. We point here to a void in discipline and accountability that would enable continuous energy improvement. Facility managers rightfully prioritize activities that make products and get them out the door. But this is too often at the expense of internal efficiencies (including energy consumption) that can potentially add to operating income. After all, a dollar saved is no less valuable than a dollar earned. People in operations, maintenance, procurement, and finance all play a role in the continuous process of energy cost control. It should not, and cannot, be a task addressed solely by the episodic engineering project.

There is also a classic perception of project economics. It goes something like this:

1. Energy projects require capital investment;
2. Project approvals should hinge on financial payback projections; and
3. Energy projects often do not provide a rate of return that is competitive to core business investment opportunities.

Why should energy efficiency be more than a project? To answer that question, think about the advice given by a good financial advisor. The advisor would first ask you to set your financial goals. You start by getting your spending under control. You would then build a portfolio of diversified securities, mixing growth-oriented equities with stable, income-producing bonds. To reach your goal, you would develop a calendar for regular contributions to your plan. Quarterly performance summaries demonstrate your current rates of return. Your investment risk is mitigated by the diversification of assets and the discipline of "dollar cost averaging" that requires consistent contributions over time.

Important note: "energy portfolio" as used here does not refer to a procurement portfolio of fuel and power purchase contracts. Instead, it refers to a collection of activities and investments that either reduce energy waste or redirect energy into more productive uses. While energy procurement portfolios are fairly common, forward-thinking organizations also use a multi-year portfolio strategy for planning energy improvement activities and investments.

These organizations start with an energy audit that compares current energy consumption patterns to optimal levels (see page 71). This information allows energy performance goals to be set. Next, they boost energy awareness by training their staff to ensure that energy-smart decisions are part of standard operating procedures (to get spending under control). Then, they develop strong portfolios that combine capital projects (equities) with best-practice maintenance discipline (bonds). They maintain momentum (make regular contributions over time) by regularly communicating results to top management and staff to sustain their support. Performance metrics and accountabilities (documented rates of return) keep people focused on results. In sum, this is how organizations create a durable, risk-managed program for continuous energy improvement. [2]

An energy management portfolio—like any financial portfolio—derives its value from the mix of its components. The logic for this is well-stated by Steve Schultz, leader of 3M's energy management program:

> *If you don't build awareness and pay attention to what you're doing every day, spending all the money in the world won't bring you the energy savings you want.*[3]

To rely on any one component in isolation is to accept tremendous risk. This is why an investment in people skills and energy-smart procedures effectively offsets the risk of capital projects. By having a strong foundation of energy-smart skills and procedures—as well as the performance metrics that provide a pulse on energy use—an organization is better prepared to implement new equipment.

Energy management is as much a communications effort as it is an engineering pursuit. The effort is ongoing because technologies evolve, labor turns over, and production costs change. Needless to say, someone needs to coordinate this effort. This is essentially the role of an energy manager, the steward of an energy portfolio that builds organizational wealth (see page 112).

Organizational Attributes, Strategies, and Outcomes

Energy solutions do not implement themselves. Even if a memo from the corporate office declares that "we shall now be energy efficient," this doesn't mean it will happen. The term "energy efficiency" probably has different meanings to people throughout the organization. People can't implement what they don't understand. Staff are ultimately motivated by the formal accountabilities set forth in position descriptions that were devised long before anyone dreamed of energy-related liabilities. If people are deeply vested in the status quo ("the way we've always done it"), they perceive change as a threat to be resisted.

Successful energy cost control is defined as much by its methods as it is by the outcomes it provides. Every company must develop a strategy that recognizes its organizational constraints, aspirations, and business culture. Organizational dynamics are those that shape internal priorities, communications, administrative procedures, and individual accountability. These dynamics determine how the concept of energy efficiency—what it means, what it entails, and what it provides—is understood at all levels of the organization. Organizational dynamics can either help or hinder efforts to control energy costs.

Research by the Alliance to Save Energy reveals five basic approaches to industrial cost control. [4] Many of these approaches are not mutually exclusive:

Do nothing. Ignore energy improvement. Just pay the bill on time. Energy initiatives are typically "crisis management," in the form of emergency repairs made without proper consideration of the true costs and long-term impacts. This strategy is pursued by companies that do not understand that energy management is a strategy for boosting productivity and creating value, or have management in such turmoil that energy management cannot be sufficiently supported, or are extremely profitable and don't consider energy costs to be a problem. Pros: Manufacturers don't have to change behavior or put any time or money into energy management. Cons: Savings are forfeited. Income is increasingly lost to uncontrolled waste.

Price shopping. Switch fuels or shop for lowest fuel prices. Make no effort to upgrade or improve equipment. Make no effort to add energy-smart behavior to standard operating procedures. Companies take this approach because they "don't have time" or "don't have the money" to pursue improvement projects. It is also preferred by companies that truly believe that price is the only variable in controlling energy expense. Pros: Management doesn't have to bother plant staff with behavioral changes or create any more work in the form of data collection and analysis. Cons: Lack of energy consumption knowledge exposes the manufacturer to a variety of energy market risks. The origin of waste is unknown, as are the opportunities to boost savings and productivity. The business remains exposed to energy market volatility. The emissions and safety compliance dimensions of energy use are not addressed.

Occasional low-cost, non-capital projects. Make a one-time effort to tune up current equipment, fix leaks, clean heat exchangers, etc. Avoid capital investments. Revert to business-as-usual behavior after one-time projects are completed. Companies that do this are insufficiently organized to initiate procedural changes or invest in assets outside of core process activities. They cannot assign roles and accountabilities for pursuing ongoing energy management. Pros: Very little money is spent when just pursuing quick, easy projects. Cons: Savings are modest and temporary because facilities don't develop procedures for sustaining and replicating improvements. Familiar energy problems begin to reappear. Energy bills begin to creep back up.

Capital projects. Acquire big-ticket assets that bring strategic cost savings. But beyond that, daily procedures and behavior remain business-as-usual. This strategy is adopted by companies that believe that advanced hardware is the only way to obtain real, measurable savings. Similarly, they believe that operational and behavioral savings are weak and not measurable. Such companies may also lack the ability or willingness to perform energy monitoring, benchmarking, remediation and replication as a part of day-to-day work. However, they have the fiscal flexibility to acquire strategic assets that boost productivity and energy savings. Pros: Obtain fair to good savings without having to change behavior or organize a lot of people. Cons: Return on investment from the new assets may be at risk if not complemented by the appropriate skilled maintenance and operating procedures.

Sustained energy management. Merge energy management with standard operating procedures. Diagnose improvement opportunities and pursue these in stages. Procedures and performance metrics drive improvement cycles over time. Manufacturers with corporate commitment to continuous improvement can pursue this strategy. They have well-established engineering and internal communications protocols and an energy program that engages staff with roles and accountabilities. They encourage cooperation among departments. Pros: Maximize savings and capacity utilization. Increased knowledge of in-plant energy use is a hedge against operating risks. Greater use of operating metrics will also improve productivity and scrap rates while reducing idle resource costs. Cons: This won't work in facilities that don't easily collaborate across departmental lines. For this to work, it is crucial that all stakeholders, especially top managers, understand and support the idea of energy management as a continuous improvement process.

The choice of strategy depends on the strength of leadership, depth of skills, and the commitment of senior management in supporting energy improvements. Table 5-1 compares the pros and cons of the five energy management strategies described above.

An organization cannot manage energy use until it identifies an implementation strategy that makes sense for its business style and attributes. Specifically, energy management capability reflects the attributes of an organization's *people* and management *systems*. A brilliant presentation by Mike White of Sunoco Chemicals describes this concept:

Strong "People" attributes:

- Strong technical skills in engineering, finance, and data management

Table 5-1. Industrial Energy Management Strategies

APPROACH	POTENTIAL FOR REDUCING WASTE	MAGNITUDE OF SAVINGS	DURABILITY OF SAVINGS	EASE OF IMPLEMENTATION	RISK OF NOT SOLVING ENERGY COST PROBLEMS
Do Nothing	None	n.a.	n.a.	Good	High
Fuel Switch / Price Shop	None	Fair	Poor to Fair[1]	Good	Fair to High
Easy, Low-Tech Projects	Fair	Fair	Poor to Fair[2]	Good	Fair to High
Advanced Tech. Capital Projects[3]	Fair to Good	Fair to Good	Good	Fair to Good	Fair[4]
Continuous Energy Improvement[5]	Good	Good	Good	Fair	Low

NOTES
1. Expect this to be an on-going responsibility if savings are to be maintained.
2. Durability is much improved if staff skills and procedures are enhanced to complement the new technology.
3. Implementation is easier if ready financing is available.
4. Risk is lower if behaviors and procedures evolve with technology.
5. Combines (1) price shopping, (2) easy, low-tech projects, and (3) advanced technology capital projects in a business-oriented management plan. Business plans typically pace implementation so that quick, easy stages come first, generating savings to pay for subsequent stages, including improvements that have a longer payback but much larger returns. Implementation requires dedicated time and skills, as well as cooperation across facility departments.

- A can-do work ethic with the ability and willingness to learn and change
- An ability to communicate work results in business terms
- Balanced respect and expectations between management and staff

Strong "Systems" attributes:

- Multi-year, organization-wide planning discipline
- Performance benchmarks, goals, and staff accountabilities
- Focused, stable leadership support for goal attainment
- Information systems for collecting and interpreting performance metrics
- Ability to coordinate goals and optimize fiscal decisions across departmental lines

Taken together, "people" and "systems" attributes determine an organization's operational style, as well as its probable approach to energy management:

Weak People Attributes, Weak Management Systems

This organization is ruled by chaos. There is uncertain leadership focus and/or control. Management has questionable abilities to detect and react to changes in the business environment. As for an energy cost control strategy, management's best option may be to do nothing, since the organization's fundamental viability as a business should demand all of the executive leadership's attention.

Weak People Attributes, Strong Management Systems

This is the quintessential bureaucracy, in which the "rules of the game" take precedence over outcomes. The current way is the best way to do anything. Without checks and balances, people can manipulate statistics to make results look better on paper than in reality. A common tactic is to delay or defer action so that others have to deal with it. In this environment, it is preferable to declare that "we're already as efficient as we can be," since attempting real change will cause unbearable friction. The best strategy for energy management may simply be to shop for the lowest available fuel and power prices—a solution that will probably resonate well with non-technical business leaders.

Strong People Attributes, Weak Management Systems

Without strong leadership and vision, the organization's agenda is driven by "fire drills," in that it is more reactive than proactive. Problem detection, definition, and solutions are local efforts with little or no coordination across departments or facilities. Strong, influential individuals are the key to problem-solving, but they tend to optimize results for their department, if not for the organization as a whole. If capital is available, this organization may rely on hardware to provide energy solutions—a series of capital projects that allow operators to flick a switch and get back to business-as-usual. If capital is scarce, the best approach may be to pursue quick, easy projects that are expensed through operating budgets. Ideally, there should be an effort to document best practices and to commit these to standard operating procedures.

Strong People Attributes, Strong Management Systems

This is the position of organizational excellence. Top management will demand and support departmental collaboration in the pursuit of continual improvement of the organization as a whole. For energy, this means benchmarking and inventorying energy use, implementing multi-year business plans for action, setting goals, accountabilities, and incentives for action, documenting and replicating lessons-learned, and demonstrating performance progress.

Strategies determine the quality of energy management outcomes. In general, strong "people" attributes allow an organization to at least initiate energy projects on an episodic basis. Strong "systems" and people attributes combined allow energy efficiency to become a durable management process that supports core business performance.

To some parts of an industrial organization, energy efficiency and its outcomes may be perceived as a benefit. Others may see it as a burden or distraction. But if facility-wide energy efficiency is to be achieved, it will require the participation of staff from operations, engineering, maintenance, and finance as well as corporate or facility top leadership. These individuals need to establish a common understanding of energy efficiency and what to expect from it if they hope to effectively control their energy costs.

In sum, these are the organizational attributes of companies that successfully manage energy costs:

- **Fundamental business viability**. The manufacturer's front office stability is important. Companies that are the subject of a merger or acquisition, labor disputes, bankruptcy or severe retrenchment may have fundamental distractions that will interfere with the attention that energy management deserves. A preponderance of such conditions indicates management turmoil that makes energy management impractical.

- **Ability to learn, document and replicate**. Companies should spread knowledge of energy efficient techniques across their multiple facilities and compare their ongoing results. The ability to cooperate—across multiple sites and across departmental boundaries—is required to maximize industrial energy management potential.

- **Energy leadership (or "champion")**. Successful energy improvements are usually led by an "energy champion," a manager that understands both engineering and financial principles, communicates effectively both on the plant floor and in the boardroom, and is empowered to give direction and monitor results.

- **Willingness to purchase energy in the open market**. This dimension is straightforward: Does the facility wish to purchase energy through open-market activity, or just procure as usual from the local utility? If open markets are the choice, the facility should be prepared to maintain sophisticated search and verification procedures to support its contracting activities (see Appendix I).

- **Leadership intensity**. Quality of operations should be demanded, facilitated and recognized by top officers of the corporation. Adoption of professional and industry quality standards, such as ISO 9000, ISO 14000, [5] or the emerging "Superior Energy Performance" energy standard ISO 50001, [6] are helpful in attaining this attribute. Energy-smart operations will hold employees accountable for adherence to energy management goals and other quality standards.

- **Positive and productive staff**. Energy efficiency is very much dependent on the behavior of line workers. Employee awareness of their impact on energy costs must be achieved. A positive, can-do attitude on the part of staff is helpful in attaining potential energy savings. Rewards and recognition can be harnessed to good effect.

- **Criteria for fiscal decisions**. Financial considerations involve far more than up-front costs. Are total costs of ownership considered? Which department pays for improvements and which claims the savings? Do savings count only fuel bill impacts or include the value of material waste minimization and greater capacity utilization? What criteria determine acceptable financial performance? Also, equipment purchasing decisions should reflect the collaboration of procurement, production and maintenance personnel.

- **Strength of engineering discipline and procedures**. Successful energy management depends on an ability to understand energy consumption. This requires benchmarking, documenting, comparing, remediating and duplicating successful improvements. To accomplish this, internal skills, procedures and information services are engaged. The likelihood of building value through energy efficiency varies directly with the depth of these technical capabilities.

Without these attributes, a manufacturer's energy management will be less effective. Or worse, the company will experience false starts and disappointing results that will bias management against future energy management efforts.

Risk, Time, and Money: The Executive Energy Tool Kit

In January 2004, the U.S. Department of Commerce issued "Manufacturing in America," a report describing challenges and threats to U.S. industrial competitiveness. [7] The Commerce Department's report describes industrial challenges that can be characterized here by three basic dimensions:

- **Risk**. Rapid changes in technology, markets, and costs of production force corporations to make money as quickly as possible, before changes in the business climate render their current activities unprofitable.

- **Time**. Publicly held companies tend to hold executives accountable for quarterly results. Therefore, many corporate leaders will avoid investments that take "too long" to generate an economic return.

- **Money**. Executives juggle the forces of risk and time to make the most effective return on investment capital.

Industry personnel working in production facilities deal with the same three forces:

- **Risk**. There is a certain amount of safety in doing things "the way they've always been done." New procedures, responsibilities, or communications often imply risk, either real or perceived.

- **Time**. Cost pressures have led to drastic reductions in facility personnel. Remaining staff handle multiple tasks and usually experience a chronic shortage of time.

- **Money**. Department heads often compete against each other for budget dollars. Energy efficiency is not welcome when the cost of attaining it accrues to one department while the savings accrue to another.

Energy efficiency in the industrial sector must somehow coexist with these challenges. Successful promotion of energy efficiency hinges on whether it is perceived as a solution to these challenges, or as part of the problem.

Any business solution is derived from a mix of three resources: risk, time, and money. You can minimize the use of any two out of three, but it will be at the expense of the third. For example:

- **Minimize your investment of time and money**. This means you assume greater risk by simply not dealing with the root causes of energy waste and volatile energy prices. It means doing little, cheap, one-time projects, and switching fuels or fuel suppliers. In other words, by doing as little proactive energy management on your part as possible, you remain at the mercy of market forces external to your organization.

- **Minimize your risk and investment of time**. You can do this primarily by pursuing big, capital projects (assuming that new equipment can do the work so people don't have to). But of course, the big project approach takes big money.

- **Minimize your risk and investment of money**. You can do this if you are prepared to invest a lot of time. If for some reason the budget won't support the purchase of new, more efficient equipment, then you need to focus on the way people use and maintain current equipment. In short, this approach requires culture change. Be prepared to spend a lot of time boosting staff awareness of the energy cost consequences of their daily work habits. Be prepared to encounter resistance ("But that's the way we've always done it.") You'll need to persuade and influence people, fostering and promoting success stories whenever they can be cultivated.

Risk, time, and money—you can optimize two out of three. This book offers no scientific proof to back up this concept, but simply asks you to accept it as a truth that is self-evident.

Energy Management: In-house or Outsource?

When seeking ways to reduce energy costs, where do business organizations go for help? For many, the choice boils down to in-house versus outsourced expertise. The issues here are many: Who provides the best outcome? Who brings truly valuable expertise to the task? Who has ulterior motives or hidden agendas that may lead to a less-than optimal solution from a business standpoint? Let's take a look at the risks involved with using in-house versus outsourced expertise.

ABOUT OUTSOURCING

Facilities may outsource energy performance analysis, system design and engineering, retrofits, construction, and energy management strategy development. By outsourcing, one takes advantage of the depth of expertise that an energy expert provides. Well-seasoned energy experts see the best and worst of many facilities, bringing a wealth of knowledge to the client. However, this expertise costs money, and there are many facility managers who cringe at the thought of writing a check to pay for a service that maybe—just maybe—could have been accomplished with in-house resources. The cost of outsourced help may be higher because of the profit margin reflected in the vendor's cost. The expert may cost more, but he might also make the solution (and therefore the savings) available more quickly. Then there may be some risk, real or perceived, that the vendor will make prescriptions that benefit equipment suppliers more than the customer who purchases the solutions. Facility managers are all wary of the "snake oil" that has been hustled by more than one energy solution provider over the years.

ABOUT IN-HOUSE EXPERTISE

There may be a sense of pride in ownership, articulated as "this is our plant, we've operated it for years and no outsider can know any better than us how to run it." A grey-beard supervisor may be the best source of cumulative knowledge about a facility, its design, and it standard operating procedures. You cannot seek or implement energy improvements without the input of such talent. On the other hand, a reliance on in-house expertise has risks of its own. Keep in mind that in today's economic environment, facility staff are usually trimmed to the bone. People "wear many hats," scrambling to keep up with multiple tasks while not mastering any one very well. They see the same workplace each and every day, not getting exposure to the lessons-learned from a variety of facilities. **Intentionally or not, in-house staff can mislead their management just as much an outside vendor**. All too often, in-house staff resist help from outside sources, primarily out of fear of embarrassment, or worse. Controlling energy costs usually involves some kind of change. Facility workers are prone to resist the "extra work" that energy optimization may entail, especially when their paycheck stays the same, regardless of energy performance. Rather than risk the exposure of waste attributable to "the way we've always done it," many in-house staff can and do advise their management not to seek outside help. Many top managers lack the technical credentials to appropriately judge such opportunities. When this happens, organizations continue to waste energy—and pay dearly for fear and misplaced pride.

The biggest cost may be the cost of doing nothing (see page 134). This is the very common result when organizations lack clear lines of "ownership" and accountability for energy issues. Unless someone is directly accountable for energy cost control, waste can go unchecked, with losses accruing not to any one department, but to the organizational bottom line.

Adventures in Energy Management: Ten Case Studies

A few energy-savvy companies have allowed their energy management experience to be documented for industry's wider benefit. No one company or industry dominates the practice. While it is easier to identify energy management leaders among Fortune 500 companies, there are also small, privately held companies that excel at stewardship of energy and other resources. It is clear that each corporation approaches energy management with a strategy that reflects the company's organizational characteristics and culture. Among the enabling determinants are the degree of corporate authority and involvement, depth of in-company technical support, leadership, and the capability to express energy performance's contribution to business goals.

Here is an overview of ten companies' accomplishments, per a study compiled by the Alliance to Save Energy in 2005: [8]

3M. This diversified manufacturer sought to reduce energy consumed (Btus) per pound of product by 20 percent over the 2000-2005 time frame. This goal required 3M's tier-1 plants (52 facilities worldwide) to achieve 3M's own "World Class" energy management label. 3M surpassed that target in under five years and uses its energy performance in its product marketing. Superior energy cost control at 3M reduces the embedded energy costs that 3M's customers would normally absorb (see page 28). Notable feature: 3M's executive management believes that resource stewardship makes good business sense. Energy management goals and results are routinely communicated to Wall Street analysts. 3M, and the manufacturers that purchase inputs from 3M, are responding to markets that increasingly demand products with low environmental impacts.

- **Authority**: Broad corporate goal to reduce overall energy consumed per volume of product.
- **Leadership**: Corporate leaders regularly review all plants' energy performance.
- **Who decides to act?** Plant managers act, with influence from plant-based energy teams.
- **Organization**: A corporate energy management team provides technical assistance and evaluation. Plant-based energy teams pursue implementation.
- **Accountability**: Energy stewardship is one of many variables used to evaluate plant performance annually.

C&A Floorcoverings. Based in Georgia, this privately held, five-plant company demonstrates successful energy management by a mid-sized manufacturer. C&A has implemented a management system for matching energy-efficiency initiatives with business goals. After two years, C&A achieved 10 percent savings on an annual natural gas expenditure of $824,500. Notable feature: C&A adopted MSE 2000/2005, an ANSI-certified standard for energy management developed by Georgia Tech, as a template for an in-house energy management program. By the end of 2004, C&A was close to becoming the first organization to become fully certified per the MSE 2000/2005 standard.

- **Authority**: Top management periodically reviews energy performance.

- **Leadership**: An energy coordinator leads all functions required by the MSE 2000/2005 standard for energy management. Top management stands behind this standard.

- **Who decides to act?** Key individuals decide to act per their accountabilities set forth in the MSE 2000/2005 standard.

- **Organization**: An in-house, cross-disciplinary team was assembled to initiate MSE 2000/2005.

- **Accountability**: Once implemented, the MSE 2000/2005 standard sets roles and accountabilities for key personnel.

Continental Tire North America. Continental began shutting down certain North American facilities due to energy waste and other cost inefficiencies. One Illinois-based facility became proactive at energy management and was rewarded by getting a larger share of overall production quotas. The Illinois plant used a combination of energy consultants and in-house management teams to achieve a 31 percent reduction in energy consumption per tire. Notable feature: Continental successfully partnered with an energy services company (ESCo) to design and implement energy management procedures that were self-sustaining after the ESCo's tenure concluded.

- **Authority & Leadership**: A facilities engineer takes nominal leadership of an in-plant energy team. Key supervisory engineers enforce energy discipline largely through personal influence and leadership. Corporate officers have no role in goal-setting or progress reviews.

- **Who decides to act?** The facilities engineer acts on the consensus of the in-plant energy management team.

- **Organization**: A cross-disciplinary energy management team discovers, evaluates, and prioritizes energy improvement opportunities.

- **Accountability**: Plant personnel generally observe in-plant leadership. While corporate officers play no day-to-day role in energy management, their long-term decisions regarding plant closure usually include energy cost performance.

DuPont. With over 100 plants in 70 countries, energy management practices at Dupont are supported by two top-level strategies. The first is designating energy efficiency as a high-priority corporate issue. The other is the application of "Six Sigma" methodology to the energy management process. Notable feature: through 2002, Dupont applied Six Sigma to behavioral tasks, including plant utility management. Over 75 energy improvement projects, many requiring no capital, were implemented across the company's global operations. The average project netted over $250,000 in annual savings.

- **Authority**: Broad, five-year, corporate-wide goals require reduced energy consumption, increased use of renewable energy, and reduced carbon emissions.

- **Leadership**: Corporate direction requires use of Six Sigma quality control methodologies for virtually all procedures at Dupont.

- **Who decides to act?** A Six Sigma culture at Dupont is the incentive for all staff to seek improvement projects.

- Organization: A corporate energy management team helps plants by providing technical assistance, documentation, and communication to build and replicate knowledge of energy solutions.

- **Accountability**: Personnel promotions at Dupont are contingent upon gaining proficiency in Six Sigma. This drives Dupont's professionals—including energy utility engineers—to improve operations through applications of Six Sigma.

Frito-Lay. This leading snack food manufacturer's energy management features aggressive energy reduction goals with a focus on results. This demands a high degree of monitoring, measurement, and communications. Frito-Lay organized the required engineering talent as its Resource Conservation Group. While surpassing intermediate targets on the way to even larger savings, Frito-Lay's efficiency initiatives have returned over 30 percent on investment. Notable feature: Large and challenging energy reduction goals were used to rally and motivate staff to generate results.

- **Authority**: Aggressive corporate goals specify reductions in energy and water. Goals are pro-rated across plants. A senior vice president for operations reviews comparisons of plants' progress.

- **Leadership**: A group leader for energy & utilities coordinates corporate-wide discovery and evaluation of improvement opportunities.

- **Who decides to act?** Plant managers and personnel make implementation decisions.

- **Organization**: Several tiers of energy leadership are involved: a corporate tier provides technical assistance; a regional tier coordinates audit functions; and site champions assume implementation details.

- **Accountability**: Corporate comparison of plants' budget-to-actual energy performance is the mechanism for ensuring compliance.

Kimberly-Clark Corporation. This personal care products manufacturer has a broad mandate for environmental stewardship. KCC's more than 165 plants worldwide practice energy efficiency, air emissions abatement, wastewater treatment upgrades, process water use reduction, packaging reduction, landfill elimination, toxic chemical elimination and environmental management system implementation. Five-year plans help coordinate benchmarking efforts across a global facility network. KCC's energy conservation efforts pursue a series of five-year plans, which seeks to expand on the success of the first plan (1995-2000). The first plan led to a corporate-wide, 11.7 percent reduction in energy use per ton of product. Notable feature: A large, global population of mills allowed KCC to generate its own proprietary energy benchmarking discipline. Sharing best practices across plants prevents "reinventing the wheel."

- **Authority**: Corporate-wide five-year plans impose goals for energy savings.
- **Leadership**: The vice president for energy and environment ultimately leads technical support, benchmarking, databasing of results, and corporate-wide communication/promotion of success stories.
- **Who decides to act?** Individual plant managers make actual implementation decisions.
- **Organization**: A corporate energy management team provides technical support, energy auditing, benchmarking, and documentation services. Plant staff perform implementation.
- **Accountability**: Energy performance is integral to plant and plant manager performance evaluations.

Merck & Co. Inc. This pharmaceutical products and services corporation seeks to improve the productivity of existing assets while reducing energy expenses. A corporate energy program is mobilized by goals that hold site managers accountable for annual performance targets. Energy costs at manufacturing sites were on a growth-adjusted pace to be cut 22 percent during 2001-2005. This equates to at least 250,000 tons of avoided carbon emissions and 11.5 percent energy expenditure savings. Notable feature: Energy efficiency was employed to boost the production capacity of existing assets, thus avoiding the need to finance new capital assets.

- **Authority**: Five-year plans establish corporate-wide goals for energy cost reduction. The senior vice president for manufacturing monitors reported energy performance.

- **Leadership**: The senior manufacturing head and energy manager coordinates corporate-wide energy management functions. Facility representatives participate in developing a 4-point strategy for planning, reporting, best-practice identification, and awareness development.

- **Who decides to act?** Each facility's general manager makes the ultimate implementation decisions.

- **Organization**: At the corporate level, "Global Energy Management" is led by an Energy Reduction Initiative Team, which is in turn comprised of a core team (for monthly review and guidance) and an expanded team (of in-house subject experts called upon as needed). Team subcommittees each represent many functions, including engineering, benchmarking, procurement, etc. Facility representatives identify improvement opportunities for their individual sites.

- **Accountability**: Energy performance is a line-item in each general manager's performance evaluation.

Mercury Marine. This manufacturer of marine propulsion systems consolidated energy decisions under the authority of a central facilities manager (CFM) and implemented a power monitoring system that permits electricity costs to be tracked and billed to individual cost centers. Valuable energy flow data gives the CFM leverage in gaining corporate approval of energy technology upgrades. The centerpiece of these efforts in 2004 was the installation of a new, centralized compressed air system that carved roughly half a million dollars from an annual $7 million electricity bill. Notable feature: Simple and effective energy management placed the authority to make energy improvements in a single manager, assigned cost control responsibility to production units, and used information technologies to monitor energy flows and to directly bill production units for their actual energy use.

- **Authority & Leadership**: A central facilities manager assumes responsibility for all energy improvement decisions, including discovery, evaluation, and technical assistance. There are no energy–saving goals or corporate reviewers.

- **Who decides to act?** Individual unit managers make energy improvement decisions per the advice of the central facilities manager.

- **Organization**: Personnel with a variety of professional disciplines form an in-house energy management team to identify improvement opportunities and assist with implementation. Unit managers petition the central facilities manager and energy team for assistance as needed.

- **Accountability**: Unit managers have cost control responsibilities. An in-plant power metering system permits direct energy cost assignment to unit managers. Energy management is therefore integral to cost performance.

Shaw Industries. Concerted efforts to manage energy at Shaw Industries got underway in mid-2004. By primarily using the U.S. Department of Energy's plant audit methods and BestPractices reference materials, a newly-hired demand-side engineer documented potential energy savings at a rate of $1 million per month for the first six months of his tenure. Notable feature: U.S. DOE website resources were effectively adopted by in-house personnel to drive their energy auditing and remediation activities.

- **Authority**: Senior management issued a general directive to "get some energy savings."
- **Leadership**: A demand-side engineer leads a corporate Energy Management Department.
- **Who decides to act?** An individual site's plant engineer or maintenance supervisor takes responsibility for action.
- **Organization**: The six-person corporate Energy management department supports plants with energy accounting, acquisitions, monitoring, and technical assistance (auditing and evaluation).
- **Accountability**: Individual plant managers are influenced by the Energy Management Department. The demand-side engineer communicates success stories to boost awareness and encourage greater responsiveness to recommendations.

Unilever HPC. Unilever's Health and Personal Care Division's energy management program coordinates 14 facilities by combining energy use targets with an energy service outsourcing strategy. A simple budget-to-actual spreadsheet compares facilities' energy performance. Notable feature: Because its use resulted in saving $4 million on energy and another $4 million in avoided costs, the spreadsheet has captured the attention of individual facility managers as well as Unilever's board of directors.

- **Authority**: All energy management results are reviewed by a senior vice president.
- **Leadership**: The energy and environmental manager leads a corporate energy team that advises staff and energy service vendors.
- **Who decides to act?** Plant managers make the ultimate decision to implement improvements.
- **Organization**: Plant managers approve a budget that incorporates planned energy consumption. Budget input comes from various stakeholders in each plant. Energy service vendors are contracted to do much of the implementation.
- **Accountability**: Quarterly budget-to-actual energy performance comparisons hold plant managers accountable for results.

A comparison of the 10 case studies presented here suggests that industrial energy management is not prescriptive in nature. It is tempting to argue that one company's results are better than another's. Upon further thought, Company A may have already been somewhat more efficient than Company B to begin with. Company B may have a steeper slope to climb because of other demands placed on its resources. The pace and extent to which a company improves its energy performance depends on forces both internal and external to the firm. The structure of authorities within companies is a major factor. So too are market conditions and asset management strategies. Is energy management helped or hindered by corporate policies regarding investment, human resource development, and outsourcing? The answers are unique for every company.

A SNAPSHOT OF CORPORATE ENERGY MANAGEMENT IN 2009

A superlative study of energy management as practiced by 48 U.S. corporations was released in April 2009 by the Pew

Center on Global Climate Change. [9] Their survey, conducted in early 2009, provides insight on how companies from a cross-section of industry formulate and implement energy cost-control strategies. While the survey includes both manufacturing and non-manufacturing companies, the sample frame was designed to purposely focus on the results of companies that have declared a proactive focus on improved energy performance. For this reason, the survey presumably captures "best practices" as opposed to "average" practices. The survey highlights a number of lessons-learned:

- The motivations for pursuing energy management vary. The most frequent reason cited by survey respondents was to contribute to the reduction of their company's carbon footprint. This was followed by (2) need to offset rising energy prices, (3) demonstration of social responsibility, and (4) to leverage energy efficiency as a way to boost productivity innovation and growth. Additional reasons are cited.

- Corporate energy programs are led by a variety of professionals. Per this survey, plant or facility managers, environmental health/safety officers, and operations directors (when counted collectively) are more than twice as often cited as corporate energy champion, compared to a corporate officer.

- Widespread engagement of all personnel. Almost 90 percent of all companies surveyed conduct outreach to engage and inform their employees with respect to energy use and its impact on business as well as in their homes.

- Moving from reactive to proactive. To varying degrees, all survey participants anticipate increasing energy prices and the onset of climate change regulation that will impact their operations. Forty-nine percent of respondents indicate that energy performance is integral to their job performance and career advancement.

- Strategic planning. Most corporations set their own internal targets and timelines for improved energy performance. The average base year for benchmarking is 2003 while the average target year for goal achievement is 2013. The average annual energy savings goal among respondents is approximately 2.2 percent. The modal value of energy reduction goals is 25 percent savings. Forty-eight percent of respondents indicate that they are meeting or exceeding their goals.

- Spin-off impacts and benefits. Energy managers cite a variety of positive impacts from their efforts. These include increased employee engagement; better communications across business units; and support, recognition, and awards from management.

- Energy management is not without challenges. These vary across companies, but often include limited capital availability, limited management leadership and support, competing priorities and resources, and lagging momentum and employee interest.

- Networking and recognition. Most companies participate in government-sponsored support programs and reference materials, such as the U.S. EPA's Energy Star, the U.S. Department of Energy's Manufacturing Energy Consumption Survey, the Carbon Disclosure Project, and other networks.

- Lessons learned. Companies applied corrective actions as they gained experience. These included increased use of energy audits to inventory and evaluate their potential improvements, team-building to enhance accountability and effectiveness, development of employee feedback mechanisms, capital fund set-asides strictly for energy, and securing upper management endorsement.

What Does an Energy Manager Do?

An industrial energy manager's job is to protect the organization's business performance from the risks imposed by today's volatile energy markets. Depending on the scope of authority vested in such a manager, the job could range from simply administering utility bills to implementing a business plan for continuous energy improvement. The energy manager's job can have implications for the organization's on-time performance, mechanical integrity, workplace safety, emissions compliance, and the ability to make products that can be marketed as environmentally benign alternatives.

The energy manager contributes to the organization's bottom line. A facility's financial performance can be dramatically impacted by sudden swings in energy expenditures—either due to fuel price spikes or unforeseen mechanical failures. By establishing an energy manager position, a facility explicitly commits to controlling its energy flows, instead of the other way around. The energy manager's regular tasks include:

- **Managing price risk**, which requires the assembly of a procurement portfolio of fixed contracts, options on future fuel purchases, and similar derivatives (see Appendix I);
- **Energy benchmarking**, to determine normal rates of energy consumption and energy-to-product ratios;
- **Goal-setting** of plant-wide targets for reducing the volume of energy required per unit of product;
- **Monitoring and verifying** energy flows, from the point of delivery through its end-use application;
- **Repairing and correcting** systems when energy consumption data indicate a significant deviation from expected levels;

- **Training and communicating** to build organizational knowledge of energy optimization and to sustain the momentum of energy-saving initiatives;

- **Scouting** for new technologies and best practices as presented by trade press and networking forums.

The energy manager concept is relatively new, compared to operations, finance, engineering, and other traditional industrial duties. Accordingly, the energy manager's agenda requires an unprecedented kind of collaboration from these other departments. Top-level direction is typically a prerequisite for getting these other departments to recognize and support energy management functions. The energy manager needs to muster strong interpersonal and communication skills in addition to technical acumen.

Energy management will require other managers to change the way they do some of their tasks. Operations directors, for example, will participate in the design and implementation of energy performance benchmarks and improvement goals. They will also harmonize their scheduled maintenance calendars so that process equipment and supporting energy systems can be serviced simultaneously. Operations and maintenance directors will work with the energy manager to develop a training calendar and criteria so that staff can obtain energy best-practice knowledge. Finance professionals assist the energy manager by modifying production data to include metrics and ratios that describe energy use. Procurement directors will work with the energy manager to develop purchase criteria that minimize the total cost of equipment ownership. Communications staff will work with the energy manager to recognize innovators and document success stories for replication throughout the organization. Top-level direction ensures facility-wide support for energy-related goals.

How much work should be expected of an energy manager? The answer depends on how much time is allocated to the task.

The energy manager role might be only a part of one person's job. Alternatively, the full list of duties may employ more than one person. At the very least, the energy manager should be responsible for energy procurement functions, especially for facilities that make purchases in deregulated utility markets. The rationale for this is straight-forward: as long as it is in operation, the facility will always purchase energy, even if it has no plans to manage consumption and reduce its waste.

Table 5-2 offers a model position description, describing specific tasks and the suggested allotment of hours per month to those tasks. As shown here, the hours are organized as "full-time equivalents" (an FTE). In other words, one FTE is equal to the total annual hours worked by a single employee. Note that it is common for industrial facilities to outsource some of these tasks, especially energy procurement, to outside vendors.

The hours and percent allocations shown here were developed by the author for illustration purposes only. The actual energy management time requirements will vary with each facility according to its type, size, hours of operation, the complexity of the energy markets in which it participates, and the facility's management style.

Note that this is a comprehensive description of duties, and it assumes that the payroll can accommodate these hours. In reality, however, this will not always be the case. Accordingly, Table 5-2 purposely organizes energy management activities into discrete levels, suggesting the priorities that should be pursued for an energy manager with limited time. For example, if a facility that decides that it can only devote one-half of an FTE to energy management, it should expect to perform only those activities in Level One (price and procurement management, representing 0.47 FTE). One full-time energy manager (1.0 FTE) could expect to accomplish all the tasks through Level 2 and some of Level 3. Note that in Table 5-2, all the energy manager duties through Level Five account for almost two full-time equivalent positions. Readers can and should adjust this interpretation to meet the reality of their staffing situations. For example, if environmental

Table 5-2. Model for an Energy Manager Position Description

Full-Time Equivalent (FTE) Distribution	Hours /Month	FTE Distrib.	Cumulative FTE
LEVEL ONE: Price/Procurement Management			
Energy price/market monitor. Track fuel markets and prices on a daily basis. Analyze current events that may impact energy markets. Manage a portfolio of energy supply to include spot-market purchases as well as futures contracts to hedge against energy price volatility.	20	0.16	
Regulatory/tariff monitor. Monitor and analyze proposed changes to utility regulations regarding gas and electricity tariffs. Maintain a calendar of regulatory activity, especially the invitations for rate-payer testimony. Represent the company in regulatory proceedings, producing both written and spoken testimony.	10	0.08	
Contract administrator/monitor. Administer energy-related contracts with outside vendors, consultants, marketers, etc.	30	0.23	0.47
LEVEL TWO: Level One plus: Regulatory Compliance Administration			
Collect data and prepare reports as required for air/water/safety regulations. Work with experts to validate and analyze data. Monitor regulations for notices of proposed rule changes and respond on behalf of the company when regulators invite industry comment or testimony.	35	0.27	0.74
LEVEL THREE: Level Two plus: Evaluation & Budgeting			
Performance evaluation. Measure energy performance on an ongoing basis. Secure and interpret energy assessments. Design, collect, archive, and analyze data that provides a "pulse" on energy performance. Create benchmarks for normal operations, and determine the boundaries that indicate the need to intervene.	40	0.31	
Planning/budgeting. Develop a calendar for energy-related activities, including purchasing, data collection, vendor bidding, etc. Coordinate with production and maintenance managers to plan certain energy improvement tasks.	24	0.19	1.24

Table 5-2 (*Cont'd*). Model for an Energy Manager Position Description

LEVEL FOUR: Level Three plus: Project Management & Implementation			
Project management. Oversee the design and installation of energy-efficient equipment and control systems, whether performed in-house or by contractor/vendor.	20	0.16	
Implementation. Design and oversee the collection, compilation, and reporting of energy performance metrics. Coordinate with other departments to ensure that energy-smart criteria are folded into standard operating procedures. Establish a protocol for addressing performance discrepancies and for rewarding innovators.	20	0.16	1.55
LEVEL FIVE: Level Four plus: Skills & Capacity Development			
Trainer/skills coordinator. Take an inventory of skills needed to improve energy performance. Compare this periodically with the skill mix of current employees. Seek or develop training resources. Develop a training calendar and milestones for staff.	10	0.08	
Communicator/coach/cheerleader. Translate energy audits and performance data into results that are meaningful to top managers in finance and operations. Update staff, regularly describing how their efforts are contributing to energy cost-control.	12	0.09	
Emerging technology monitor. Be on the look-out for new technologies and practices that can be added to (or substitute for) current status-quo. Read trade press, participate in professional societies, attend trade shows.	5	0.04	1.77
MEETINGS & ADMINISTRATIVE			
Allow time for routine administrative tasks and standard management meetings.	24	0.19	1.95
TOTAL	250	1.95	

compliance duties are already tasked to another staff person, then that task should be extracted from the summary in Table 5-2 and the FTE calculation adjusted accordingly.

Another interpretation of Table 5-2 is that the overall energy management agenda is best pursued by a team as opposed to one person. There are several advantages to the team approach:

- The work load is distributed across a number of individuals.
- The energy team can draw its members from operations, finance, maintenance, and procurement—thus ensuring that these departments are properly represented in the energy decision-making process.
- Team membership can be revised over time to bring in new people and build energy knowledge throughout the organization.
- Energy team leadership is an opportunity to get a big-picture understanding of facility operations, and to build skills for managing project implementation, human resources, budgeting, and other activities.

While top-management backing is important, the energy manager also needs to be a strong, visionary communicator who makes the benefits of energy management clear to everyone from the boardroom to the plant floor. A solid understanding of finance and accounting principles will help greatly, although a good working relationship with the finance team will serve the same purpose. The energy manager is able and willing to keep abreast of technical developments in the field, and will take advantage of trade press and professional society membership for this purpose. In sum, it is the ability to contain business risks and capture emerging opportunities that makes the energy manager position truly effective.

Energy Management: How Well Are You Doing?

To some degree, every organization's energy performance falls short of its potential. Current performance reflects a level of attainment for each of these dimensions:

- **Organizational policy**. How well does the organization articulate its energy vision and goals?
- **Management structure**. Is there clear authority and accountability for energy-related decisions?
- **Implementation and motivation**. Is energy policy integrated with standard operating procedures?
- **Investment analysis**. Are the right criteria used to reach conclusions about energy and money?
- **Monitoring and targeting**. Is there a way to measure and react to energy performance?

Table 5-3 explains these concepts and allow readers to make a self-assessment of their organizational energy performance potential. The questions to ask are: "Where am I now?" "Where do I want to be?"

Endnotes

1. Thanks to Greg Conderacci of Good Ground Consulting for sharing this concept.
2. See page 71.
3. Personal communication, May 12, 2009
4. http://www.ase.org/content/article/detail/2866. Accessed March 2, 2009.
5. http://www.iso.org/iso/management_standards.htm. Accessed March 2, 2009.
6. http://www.superiorenergyperformance.net
7. http://www.ita.doc.gov/media/Publications/pdf/manuam0104final.pdf. Accessed March 2, 2009.
8. http://www.ase.org/section/topic/industry/corporate/cemcases. Accessed March 2, 2009.
9. http://www.pewclimate.org/energy-efficiency/survey_paper

Table 5-3. Where Am I Now, and Where Do I Want to Be?

LEVEL OF ATTAINMENT	ORGANIZATIONAL ENERGY POLICY	MANAGEMENT STRUCTURE	IMPLEMENTATION & MOTIVATION
★★★★	An energy policy is developed with annual action plan updates. Top management imposes performance goals and objectives that are integral to the organization's overall business strategy.	An energy manager has the resources and authority to implement efficiency improvements. Energy cost control is an integral part of organizational policy, planning, operations, and accountabilities.	Energy efficiency principles are fully integrated with standard operating procedures for operations, maintenance, and procurement. Employees and vendors are also accountable for these principles.
★★★	An energy policy has been developed and published, but it's really maintained for public relations purposes. It has only limited application as a management tool.	A facilities manager has been assigned energy cost control at least on a part-time basis. This person has technical abilities, but may not be prepared for financial analysis or for leading organizational change.	Implementation proceeds, but is perceived as a task that competes with "normal" activity. The energy leader may have a few designated subordinates to assist.
★★	A draft energy policy exists, but it has no serious attention or backing from senior management.	A facilities manager provides energy status reports occasionally, as requested by directors or management committees.	A facilities manager reviews energy cost awareness from time-to-time with other middle managers. Improvements tend to be projects, rarely involving changes in behavior or procedures.
★	Facilities staff pursue isolated initiatives, but without guidelines, funding, or analysis of results.	Energy management is a part-time interest for an individual who acts independently and has no authority to ensure the cooperation of others.	Energy cost control is strictly a boiler room activity. Behaviors and procedures for everyone else are all business-as-usual.
NO STARS	No explicit policy or concern about energy waste.	No one actively tries to reduce energy waste.	Energy cost control is not discussed. No activities are planned or underway.

Table 5-3 (Cont'd). Where Am I Now, and Where Do I Want to Be?

INVESTMENT ANALYSIS	MONITORING & TARGETING	OUTCOMES
Energy is managed just like wealth. Performance goals and metrics apply to energy consumption. An energy management portfolio directs investment in energy improvements. Pre-determined investment criteria reduce the time and money lost to pondering.	An energy management information system monitors consumption, identifies discrepancies, quantifies dollars at-risk, tracks actual-to-budget performance, and calculates scorecards of performance.	The organization manages its energy costs within targets. It knows how much energy it uses, and at what cost. It knows what improvements to make and when to make them.
Energy efficiency investments compete against non-energy investment opportunities in the capital budget process. Evaluation depends on simple payback criteria.	Energy use is sub-metered within facilities. Energy-use data is distributed among a few key technical staff, but not always to other staff.	Energy cost control is a project-focused pursuit. The pace of implementation depends on the persuasion skills of key personnel.
Successful energy savings initiatives await budget surpluses if they are to happen at all. Alternatively, energy performance contractors are compensated through the shared savings from the projects they implement.	Cost monitoring and reporting is limited to utility bills, with little or no ability to disaggregate the energy cost of distinct zones or departments behind the meter.	Energy projects are justified one at a time. Implementation is delayed and money wasted because of the time needed to justify and ponder each project proposal. No capacity to measure before-and-after impacts.
Obvious low- and no-cost initiatives are pursued sporadically, as time, money, and interest permit.	Current utility bills are compared to those in recent memory.	Small and temporary energy savings. Quick and easy solutions address symptoms, not root causes.
Energy cost control is focused entirely on the purchase price of energy. Investment is driven by replacement needs, not efficiency.	Energy cost performance is not tracked. Just pay the bill on time.	The organization is at the mercy of volatile energy markets.

Chapter 6

How Does The Money Work?

"Spare no expense to save money on this one."
—Samuel Goldwyn
(1890-1969)
U.S. movie producer

Despite a lack of formal education and a ruthless business style that created many enemies, Goldwyn nevertheless prospered in a highly competitive business by consistently emphasizing quality products.

Energy At-risk: Save or Buy?

Manufacturing facilities frequently have the opportunity to make certain inputs in-house or to buy them from outside suppliers. In the context of energy use, the make-or-buy concept directly applies to those facilities that enjoy the ability to generate electricity onsite instead of having it supplied by a utility. The same concept, with slight modification, describes the opportunity to either save or buy a unit of energy. As this section shows, the save-or-buy criterion can be an effective decision tool for selecting energy cost control initiatives.

As long as a business operates, it commits to consuming energy. Consider this:

- Energy consumption can be divided into two proportions: that which is committed versus that which is wasted.

- *Committed energy* represents that which is applied as intended plus some increment of loss that is not yet technically or economically recoverable.

- Energy waste that can be economically avoided is *energy at-risk.*

The facility WILL PAY for energy at-risk, by either purchasing it, or by paying the cost to avoid it. This is depicted in Figure 6-1.

When considering the implementation of a specific energy improvement, the business choice is simple, as shown in the figure above: either (A) continue buying the energy at-risk at the prevailing price, or (B) implement an energy-reducing improvement when the cost to save energy on a per-unit basis is less than the price to purchase it.

Now here's what we want to do: develop a management tool for making energy cost-control decisions. This tool needs to compare the financial merit of implementing energy improvements to simply continuing to buy and waste the portion of energy at-risk as described above. Also, this tool needs to account for the organization's cost of capital and fit logically with the annual

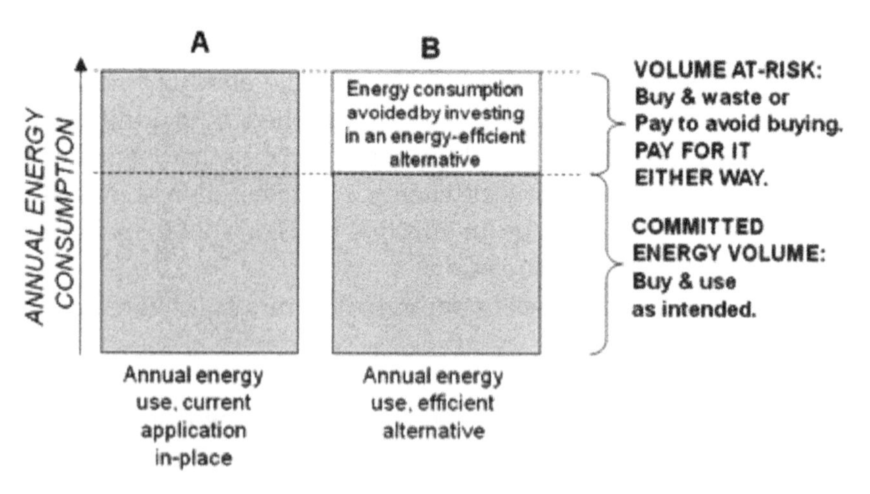

Figure 6-1. Energy At-risk

parameters that shape budget and performance accountabilities. That tool is the save-or-buy calculation.

The save-or-buy calculation requires the annualization of all relevant cash flows and investments, so that costs and benefits can be correctly compared on an "apples to apples" basis. Most financial targets, measures, and budgets are already expressed on an annual basis. The total cost of a large asset that will be in service for more than a year is typically financed over an equivalent number of years. To express the value of these investments as an annual equivalent, we use amortization—a calculation that organizes an investment's capital and interest costs in a series of annual payments of fixed amount. [1]

The cost to save a unit of energy is calculated in two steps. First, determine the total up-front cost to implement a specific energy-saving initiative. Then annualize that project cost as follows:

$$\begin{matrix}\text{Annualized}\\\text{Project Cost}\end{matrix} = \begin{bmatrix}\text{Up - front}\\\text{Project}\\\text{Cost}\end{bmatrix} \times \begin{bmatrix}\text{Capital}\\\text{Recovery}\\\text{Factor}\end{bmatrix} \qquad (Eq.\ 6\text{-}1)$$

$$\text{Capital Recovery Factor (CRF)} = \frac{i(1+i)^n}{[(1+i)^n]-1}$$

Where:

i = cost of capital or discount rate on future cash flows

n = economic life (years) of remedy (energy improvement project)

The second step is to distribute the annualized project cost over the volume of kilowatt-hours, therms, gallons of oil, or other units saved, as appropriate, that the project provides in its first year:

$$\begin{matrix}\text{Cost to Save}\\\text{A Unit of Energy}\end{matrix} = \left[\frac{\begin{matrix}\text{Annualized}\\\text{Project Cost}\end{matrix}}{\begin{matrix}\text{Units of}\\\text{Energy Saved}\end{matrix}}\right] \qquad (Eq.\ 6\text{-}2)$$

The resulting *cost to save a unit of energy* should be compared to the *delivered price* to buy that same unit of energy.

An example. A manufacturing plant contemplates replacing its current boiler and steam system. While it appears that the current system could continue functioning for the foreseeable future, its efficiency has nonetheless been compromised by age and neglect. The boiler consumes natural gas (measured in therms). The relevant investment data are shown below.

Figure 6-2

Up-front Project Cost:

Construction cost:	$239,305
Engineering fees:	$29,900
Total installed cost (TIC)	$269,205

Investment Criteria:

Current price per therm:	$1.611
Economic life of new boiler (n):	25 yrs
Discount rate/cost of capital (i)	8%
Capital recovery factor (CRF):	.0937 - $[i(1+i)^n]/[((1+i)^n)-1]$

Savings Results:

	Old	New	Savings
Therms consumed/year:	390,780	298,998	91,782
Annual fuel cost:	$629,547	$481,686	$147,861

With these inputs, we get:

$$\begin{matrix}\text{Annualized} \\ \text{Project Cost}\end{matrix} = \begin{pmatrix}\text{Up-front} \\ \text{Project} \\ \text{Cost}\end{pmatrix} \times \begin{pmatrix}\text{Capital} \\ \text{Recovery} \\ \text{Factor}\end{pmatrix}$$

$$\$24{,}225 = (\$269{,}205) \times (.0937)$$

$$\begin{array}{c}\text{Annualized}\\ \text{Project Cost}\\ \text{Per Annual}\\ \text{Therm Savings}\end{array} = \frac{\$25,225}{91,782} = \boxed{\$0.2748}$$

(*Eq.* 6-3)

In this example, the investor has two choices: continue to buy the energy at-risk (91,782 therms) at the current price of $1.611 per therm, or pay an annualized cost of $0.2748 *per therm avoided* as the result of investing in the boiler replacement. The ratio of the price to buy versus the cost to save each unit of energy at-risk provides a cost-benefit measure:

$$\frac{\begin{array}{c}\text{Cost to}\\ \text{Save a Therm}\end{array}}{\begin{array}{c}\text{Price to}\\ \text{Buy a Therm}\end{array}} = \frac{\$0.2748}{\$1.611} = 0.17$$

(*Eq.* 6-4)

Stated differently, this project would allow the investor to pay $0.17 to avoid buying a dollar's worth of energy.

Note that the *annualized cost to save a unit of energy* effectively amortizes project costs, so that the annual budgeted value remains constant over the economic life of the project. Amortized project costs can be budgeted each year with certainty. In contrast, volatile energy prices lead to volatile budget performance. By reducing energy waste, facilities reduce their budget risk accordingly.

In sum, an energy-consuming organization has one of two choices for the at-risk portion of their energy consumption. They are:

- **Buy it**. The facility chooses not to make the energy improvement. It will continue to buy more energy than it needs to accommodate its waste. For whatever reason, people are not motivated to change.

- **Save it**. Alternatively, the facility can implement efficiencies that allow the recapture of energy waste so that it can be re-applied to useful purposes. By "recapture," we mean anything that reduces the loss of energy. Recaptured energy allows the facility to offset its energy purchases by a corresponding amount.

Figure 6-3 illustrates concept of offsetting some amount of purchased energy by recapturing energy waste:

In Figure 6-3, energy consumption is broken down into segments as follows:

- A, B, C, and D, prior to any energy improvements, are all purchased at the prevailing market price.
- A and B are used as intended.
- C and D are purchased, but lost to waste.
- C represents the *energy at-risk,* that is, energy waste that can be economically avoided.
- D represents energy waste that cannot be economically avoided at the present time.

As the value in segment C is recaptured, it replaces the value in segment B. The value in segment B, net of the annualized cost of energy improvements, is extracted from the energy procurement budget altogether and is redirected into operating income.

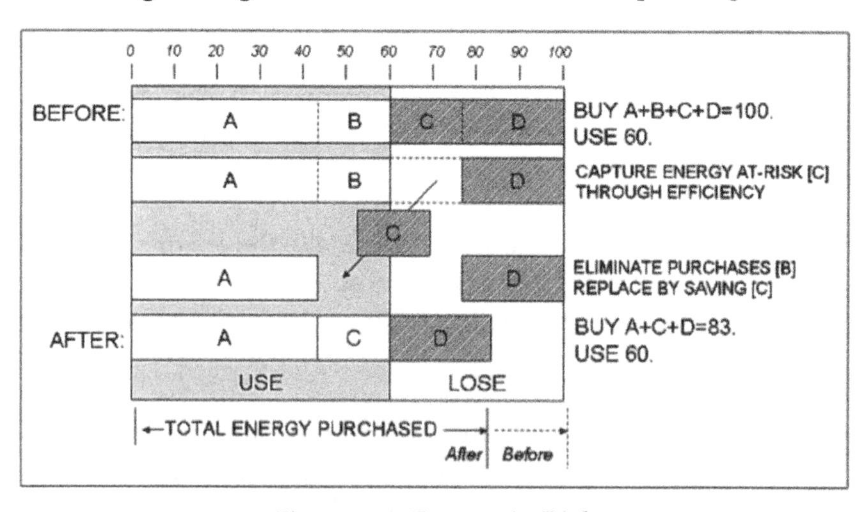

Figure 6-3. Energy At-Risk

Also, new technologies and practices emerge over time that will make it cost-effective to recapture an increasing portion of the value in segment D.

What if the organization is unable or unwilling to devote capital to the implementing the energy improvement? At the very least, this calculation still puts a dollar value on forfeited net savings—an amount that will encumber operating budgets as long as energy waste is allowed to continue. This "cost of doing nothing" is the subject of a following section.

The organization can use save-or-buy analysis to always obtain the lowest total cost of ownership for any particular energy improvement opportunity. The business challenge now is to consistently monitor the price per unit of delivered energy versus the cost of saving the same unit. Hopefully, the organization will have conducted an energy audit that identifies potential energy improvements. When it's cheaper to save a unit of energy than it is to buy it, then the save-or-buy decision should become clear.

Capital Investment, Capital Projects

Sooner or later, an organization that is serious about energy cost control will pursue **capital projects**. This typically entails investments in assets that will be financed over a number of years. These investments include key energy-using hardware such as boilers, air compressors, chillers, metering systems, etc.

When seeking results *strictly through capital projects*, here's the basic financial equation for linking the value of energy saved to the amount of invested capital:

$$\text{Capital Required} = \frac{\text{Percent Desired Reduction in Energy Expense} \times \text{Total Energy Expense}}{\text{Discount Rate or Cost of Capital (Desired Return on Investment)}} \qquad (Eq.\ 6\text{-}5)$$

You'll see this formula in real estate, for example, for determining a property's capitalized value as a function of the income it generates. In this instance, however, we are capitalizing an increment of avoided expenses—expressed here as a percent reduction in energy expenses. We can use "expenses" interchangeably with "income" because *a dollar saved is a dollar earned.*

When using this formula as written, one calculates the maximum capital to be expended (given a current level of energy expenditure), a hurdle rate, and a desired percent reduction in energy expenses. An alternative approach for the same formula is to insert a fixed amount for capital expenditure and solve instead for the desired percent reduction in energy expense. In other words, this second approach describes the percent savings that *should* be achieved by a certain amount of capital expenditure. Notice also how the volume of allowable capital expenditure decreases as perceived risk rises. The higher the risk, the higher the hurdle rate—the denominator of the equation shown above (desired return on investment).

An energy reduction strategy based solely on capital investment is susceptible in several ways. For one thing, new assets will not provide superior results if people don't use them properly. To offset the risk of failure, finance people will arbitrarily raise the discount rate for evaluating energy projects. Just how much of an adjustment is needed is a matter of individual judgment. A higher discount rate yields less capital for an energy manager to work with. The energy manager can and should offset capital investment risk, effectively lowering the discount rate. How? New capital assets should be accompanied by behavioral and procedural changes that reduce energy waste. Staff need to be aware of wasteful practices. Energy-smart behavior needs to become a part of standard operating procedures. Energy cost-control performance needs to be wrapped into daily job accountabilities. This effort requires some communication, discipline, and cross-departmental coordination. Companies that develop these management skills can effectively reduce the risk on their future capital projects.

Energy Management Without a Capital Budget

If industrial decision-makers resist capital investment to reduce their energy consumption, this really leaves one other alternative: change the way we use the same ol' machinery. Only a heretic would suggest this approach to facility with a strong engineering culture. Yet there are people in industry who are taking a bite out of energy costs with little if any capital expenditure. Their primary tools are behavioral and procedural change.

Among the most accomplished players in industrial energy management was Jim Pease of Unilever. Attached to Unilever's Home & Personal Care division, Jim was a corporate safety, health and environmental compliance advisor for 14 North American sites. In 2000, Jim was tasked with energy cost control, as were many corporate environmental managers around this time. And like many of his peers, Jim's cost control mandate came with no capital budget support.

Jim's epic story is captured in one of the energy management case studies assembled by the Alliance to Save Energy.[2] The primary tool in Jim's capital-free strategy was a spreadsheet to track actual-to-budget energy consumption, normalized for production levels. In this spreadsheet, current month results were color-coded to instantly show how well a plant is doing: red for overages in excess of 20 percent, yellow for overages 20 percent and below, and green for better-than benchmark performance. All 14 sites could be compared at a glance—a feature that often mobilized friendly competition between sites.

But what about actual energy saving measures? Jim understood that when machines ran unnecessarily, the facility was wasting money, but he needed equipment operators to be aware of that. Instead of taking an adversarial approach, Jim reinforced good energy behavior through an upbeat and sometimes humorous communications campaign. To ensure that he was the bearer of good news, Jim made sure that staff got energy information that was valuable to them at home as well as in the workplace.

Messages that were brief and positive, yet frequent, characterized his campaign.

When asked about maintaining momentum behind energy management efforts, Jim's response was simple: keep it fun. He approaches staff not as a threatening reaper, but as a friendly resource who gives employees energy-saving info to take home to their families. This is part of an effort to create a round-the-clock awareness of energy waste that ultimately reinforces energy-smart behavior in the workplace.

Of course, only so much can be accomplished through behavioral change. But by tracking results and creating a buzz about early results, Jim paved the way for capital investment in subsequent rounds of energy cost control.

Simple Payback: Wrong Tool for the Task?

Industrial decision-makers everywhere depend on "payback" as a way to evaluate proposed investments in their facilities. Compared to more sophisticated financial measures such as net present value and internal rate of return, payback is comparatively simple to understand and calculate—perfect for back-of-the-envelope analysis. But its inherent simplicity also creates problems. As a managerial decision tool, payback remains grossly inexact and misapplied, especially when thousands or even millions of dollars are at stake.

Simple payback, of course, is a measure that describes the number of years that it takes for an investment to pay for itself through the annual savings or benefits that the investment creates. To calculate it, one merely divides the total cost of a proposed investment by the annualized savings (or benefits) that the investment will provide.

Let's consider the proper—or improper—use of simple payback. People frequently fail to account for all eligible costs and benefits in their payback analysis. *Project cost* may be described simply as the "catalogue price" for the equipment in question. But when you think about it, a discrete project incurs a number of ancillary costs. These may include:

- Search and evaluation costs
- Consultant fees
- Sales commissions
- Permitting or construction fees
- Installation fees
- Removal/scrap of old equipment
- Finance transaction costs
- Revenue lost to downtime during installation of the energy improvement
- Net projected salvage value of the new equipment (usually a positive value)

The same concern applies to the "annual savings" figure. When it comes to energy projects, there is a tendency to count only the energy savings generated by the new equipment. A broader definition of savings should:

- Include annual savings in energy costs
- Subtract costs of upkeep
- Subtract changes in other operations and maintenance expenses
- Subtract monthly finance charges
- Add any non-energy improvements, such as reduced waste of water, raw materials, labor, etc.

Another problem with payback is the conceptual "blinders" worn by its users. By this, we mean the fact that managers will commit to memory the payback that was calculated at a specific point in time for a project proposal, for example: "That boiler upgrade is a five-year payback." Let's say that result was true in 2002 when natural gas cost $3.00 per MMBtu. What was the payback in 2008 when gas prices topped $14.00?

Similarly, interest rates vary every day, and as a result, so does the organization's cost of capital. Calculations like the annualized cost to save a unit of energy vary accordingly (see page 121). However, most organizations do not allow their payback criteria to vary with interest rates. Why? As this section explains, payback is a risk management tool, not a measure of profitability. Additionally, it is a measure of time, and not the cost of money (because it ignores interest rates). Note that production targets and budget amounts are fixed in an annual format. So are performance evaluations and bonuses. This is why simple payback prevails despite its weaknesses: it fits naturally with the priorities of managers whose spending authority and performance criteria are fixed in a calendar-driven framework.

Is simple payback the right calculation for evaluating proposed energy improvements? If not, what questions should be asked about such investments, and how best should those questions be answered?

Think about why we perform financial analyses in the first place. Whenever a business invests in itself, it implies making a change. With change comes risk. Before committing to an investment, top managers will want to know the risk of losing their money, or at least the risk of failing to invest in more valuable alternatives. Energy investments, like all investments, put time and money at risk. Therefore, proposed energy improvements need to withstand hard scrutiny:

- What's the value that the investment will provide?
- How quickly will the benefits become available to the business?
- What will the proposal cost?
- What's the most that should be paid for the improvement?

- How does this investment compare with other ways to use money?

Investors use payback simply to decide whether they will accept or reject an investment proposal. The greater the investor's concern with investment loss, the shorter the payback time demanded. For example, a 12-month payback is preferred to a 24-month payback, and a 6-month payback is preferred to a 12-month payback.

Now take this to its logical conclusion: **a zero-month payback would be most preferred—because there's no wait to get the money back**! The investor is assured of avoiding loss only by making no investment at all.

Payback only indicates if the investor should part with the money. It reduces investment analysis to a "yes/no" decision. As a consequence, this approach reduces energy management to a stop-and-go process. The company's beleaguered energy manager has to reset his or her agenda back to zero with each project rejection.

Energy improvements need to be held to a different standard. Why? As seen on page 121, once a business commits to operations, it commits to using energy. The question is not *if* it will buy energy, but *how much* energy will it buy, and at what price. Energy costs are not a yes/no choice, but a question of degree. Specifically, how much do you want to pay for energy that ends up being wasted?

Let's be absolutely clear about this: **use simple payback when the alternative to investment is to keep the money**. Examples of investments like this would be to build or expand a facility, install equipment specific to a new product line, or to buy back stock from shareholders. This is not the situation for the volume of energy at-risk. The alternative to making an energy-saving investment is to continue buying the energy that will be wasted—keeping the money is not an option! Proposed energy improvements should be evaluated by comparing the *annualized cost to save a unit of energy to the delivered price of buying that same* *unit of energy.*

The take-away points of this section: [3]

1. By focusing on payback, the business explicitly perceives energy cost control as a series of isolated projects as opposed to a coordinated and continuous improvement process (more on this on page 85).

2. Payback is severely limited in its ability to analyze investment performance. Misuse of the payback concept can actually result in bad management of wealth.

3. **Make sure that your evaluation of energy projects asks the right questions—and that you use the right tool for answering those questions.** Looking for a better way? See the save-or-buy analysis on page 121.

Simple payback does nothing more than suggest how long it takes for an investment to pay for itself from the savings it provides. It cannot indicate profitability, so it is useless as a tool for comparing the financial performance of alternative investments. As this article explains, once a business decides to operate, it commits to using energy. It will either *buy and waste* the energy at-risk, or it will pay to reduce that volume of consumption. The save-or-buy criterion is the decision tool for making that choice.

The Cost of Doing Nothing

In the absence of a management standard for continuous energy improvement, individual initiatives must be identified, documented, promoted, and deliberated one at a time. Delay invokes its own cost, since energy waste waits for no one.

Let's say a facility has conducted an energy audit that identifies a list of potential improvements. However, decision-makers may disagree, not understand, or otherwise need to be assured of the proposed benefits. This means more presentations that have to be scheduled sequentially on people's calendars. It may take the organization months to ponder a single proposal. Meanwhile, the clock is ticking, and energy is still being wasted.

What is the cost of not making an energy improvement? To answer this, revisit the energy at-risk concept presented on page 121. The expenditure at-risk is the dollar amount of gross energy savings potential. Savings, of course, are usually achieved at some cost, depicted here as annualized project cost. That leaves a balance of value remaining, as shown in Figure 6-4:

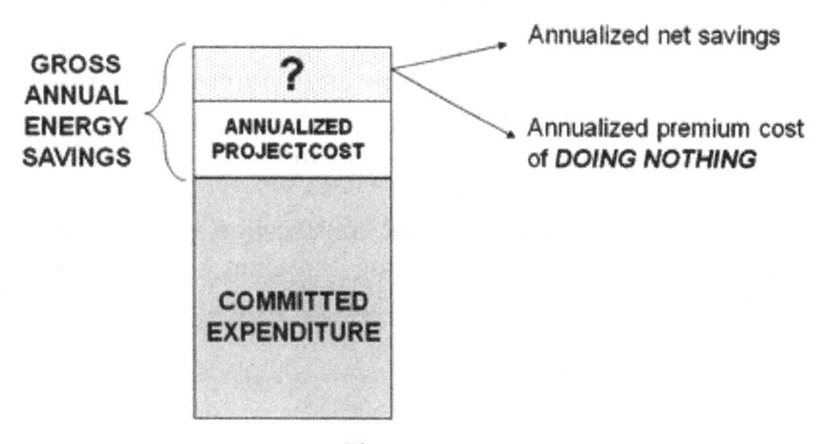

Figure 6-4.

If the organization accepts the energy improvement proposal, that balance becomes annualized net savings, which then accrue as additional operating income.

This concept permits the calculation of the *cost of doing nothing,* or in other words, the **annual premium in dollars that the organization pays because it chose to reject the net savings provided by an energy improvement.**

$$\left[\begin{array}{c}\text{Price per unit}\\\text{to buy energy}\end{array} - \begin{array}{c}\text{Annualized cost}\\\text{to avoid purchasing}\\\text{a unit of energy}\end{array}\right] \times \begin{array}{c}\text{Volume of}\\\text{avoidable}\\\text{energy}\\\text{purchases}\end{array} = \begin{array}{c}\text{Annualized}\\\text{penalty for}\\\text{Doing Nothing}\end{array}$$

(Eq. 6-6)

This formula reveals several important business implications. The *cost of doing nothing* about energy improvements goes up:

- as the price per unit to buy energy goes up,
- as interest rates go down, and
- as the price declines for steel, copper, brass, and other commodities used to fabricate new, energy-saving equipment.

To understand the second point, recall that the capital recovery factor (CRF) is used to determine annualized project costs. Interest rates are integral to CRF calculation. As interest rates go down, so do the costs of financing energy improvement projects, which yields more annualized net savings. Similarly, the price of steel, copper, brass, and other materials directly influence the cost of energy-saving capital projects.

The cost of money, fuel, and construction commodities all determine the cost of doing nothing. An energy manager needs to track these variables over time, and be prepared to act accordingly.

Break-even Cost: The Limit to Capital Investment

Another application of the energy at-risk concept (page 121) allows us to calculate the break-even amount to be invested in an energy improvement project. The industrial energy user should seek any opportunity to reduce waste when the annualized cost to save energy at-risk is less than the cost to purchase that energy at prevailing prices.

Another way to put this: with all else being equal, an investor should be willing to spend up to 99 cents to avoid the commitment of spending a dollar. At its break-even point, the cost of a solution is just equal to the benefit it provides.

The comparison calculations are as follows. Recall from page

121 the capital recovery factor (CRF), which allows:

$$\text{Annualized Project Cost} = \begin{bmatrix}\text{Up - front}\\\text{Project}\\\text{Cost}\end{bmatrix} \times \begin{bmatrix}\text{Capital}\\\text{Recovery}\\\text{Factor}\end{bmatrix} \textbf{ AND } \frac{\text{Annualized Project cost}}{\text{CRF}} = \text{Up - front Project Cost}$$

(*Eq.* 6-7)

The issue here is energy at-risk; i.e., the portion of current consumption that is potentially avoidable. The business **will spend money** on energy at-risk, either by purchasing that volume and wasting it, or by paying for a solution that eliminates the waste. The business objective is to find which approach is more valuable. To do this, the business needs to employ a decision tool that indicates a "break-even" point where the cost to *avoid purchasing* a unit of energy is just equal to the price at which the unit of energy would be purchased:

$$\text{Annual value of avoided energy purchases} = \begin{bmatrix}\text{Delivered}\\\text{price per}\\\text{unit of}\\\text{energy}\end{bmatrix} \times \begin{bmatrix}\text{Units of}\\\text{avoided}\\\text{energy}\\\text{consumption}\end{bmatrix} = \text{Maximum acceptable annualized project cost}$$

(*Eq.* 6-8)

...or more simply:

$$\text{Annual Value Of Avoided Energy Purchases} = \text{Maximum Acceptable Annualized Project Cost}$$

(*Eq.* 6-9)

AND BECAUSE:

$$\frac{\text{Annualized Project Cost}}{\text{CRF}} = \text{Up-front Project Cost}$$

THEREFORE:

$$\frac{\text{Maximum Acceptable Annualized Project Cost}}{\text{CRF}} = \text{Maximum Acceptable Up-front Project Cost}$$

(*Eq.* 6-10)

So:

$$\text{Maximum acceptable up-front project cost} = \frac{\begin{bmatrix}\text{Delivered} \\ \text{price per} \\ \text{unit of} \\ \text{energy}\end{bmatrix} \times \begin{bmatrix}\text{Units of} \\ \text{avoided} \\ \text{energy} \\ \text{consumption}\end{bmatrix}}{\text{CRF}}$$

(Eq. 6-11)

Let's apply this using the boiler replacement example described on page 121:

$$\text{Maximum acceptable up-front project cost} = \frac{\begin{bmatrix}\text{Delivered} \\ \text{price per} \\ \text{unit of} \\ \text{energy}\end{bmatrix} \times \begin{bmatrix}\text{Units of} \\ \text{avoided} \\ \text{energy} \\ \text{consumption}\end{bmatrix}}{\text{CRF}} = \text{Break-even project cost}$$

$$\text{Maximum Acceptable Up-front Project Cost} = \frac{(\$1.611) \times (91{,}782)}{0.0937} = \$1{,}578{,}383$$

NOTE: CRF = 0.937 when n = 25 and i = 8%

(Eq. 6-12)

What does this mean? Given the investor's cost of capital (8%), the economic life of the investment (25 years), and the volume of energy that it saved every year (91,782 therms), this investment would be cost-effective at any total installed cost up to $1,578,383. Recall from Section 6.1 that the total installed cost for this project was $269,205. This proposal should be a slam-dunk!

But wait—the investor would be more comfortable if assets were fully depreciated in five years. No problem. With a five-year time horizon, and with a cost of capital of eight percent, the capital recovery factor is now 0.2505. The break-even cost is then calculated to be $590,365. That's still way above the invoice cost of $269,205.

Determining a Budget for Additional Energy Analysis

The energy at-risk concept provides yet another way to organize the value of energy saved. Recall from page 134 the concept of net annualized savings, which is the balance of value that remains after annualized project costs are subtracted from gross annual fuel savings:

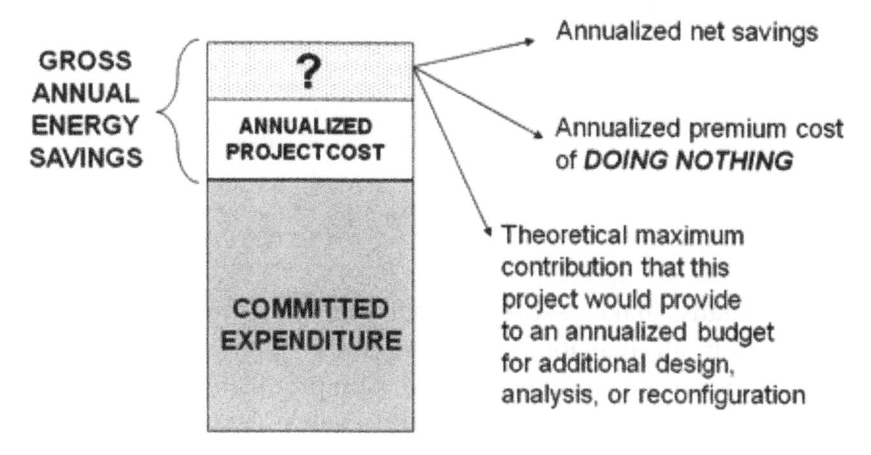

Figure 6-5

Consider a portfolio of energy improvement projects resulting from an energy audit's list of recommendations, each providing some measure of annualized net savings. The composite savings from these projects represent a new annualized cash flow. What can be done with that money? Let's say a facility has four major energy improvement opportunities. The potential value can be organized as shown in Figure 6-6:

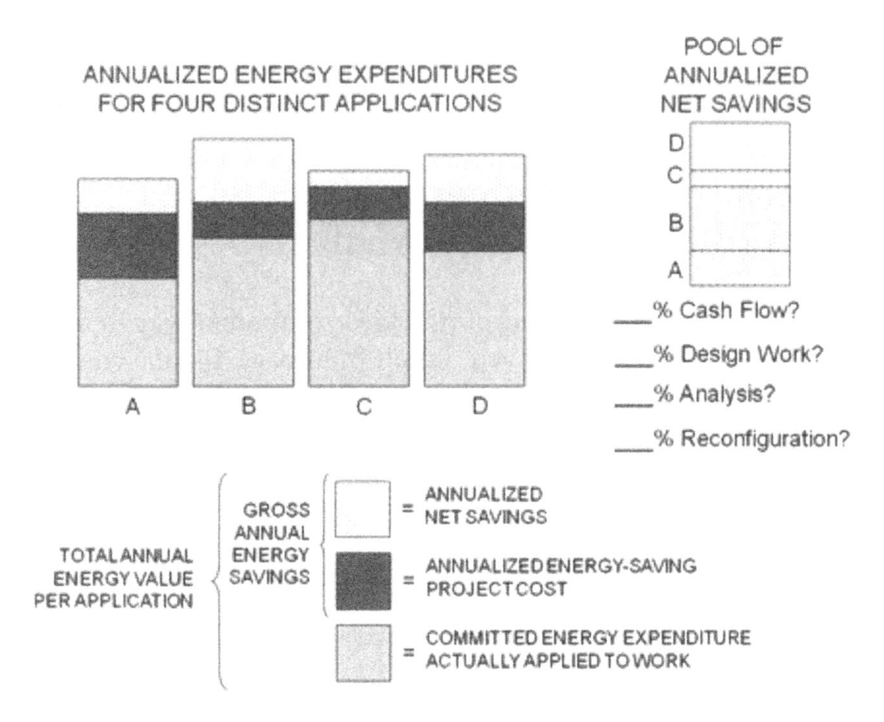

Figure 6-6

It usually makes sense to pool the net annualized energy savings from all available initiatives. Some opportunities will be more fool-proof than others, while certain promising but risky initiatives deserve additional design and analysis work. The pooled energy savings can be applied to a variety of design, analysis, or project refinement activities.

A self-sustaining energy improvement program will retain some of these pooled savings to fund recurring savings opportunities. Also, some or all of the savings can be employed immediately as working capital for the core business. The pool of savings can be allocated to a number of uses. Exactly what percent goes where is simply a discretionary choice:

- Direct some savings to earnings and shareholder wealth. A dollar saved is a dollar earned.

- Divert some proportion to other investment opportunities to save expenses, build revenues, or both (as some energy improvements will do).

- Apply some funds to the additional design, analysis, or reconfiguration of proposed energy improvements. For example, let some of the savings from the installation of efficient light fixtures pay for the implementation of occupancy sensors that will generate savings on top of the new light fixtures themselves.

Final note: if all the annualized net savings are spent entirely on design and analysis work, the energy savings will be cancelled out. It may be appropriate to apply a discretionary limit to the amount of pooled savings that are directed to a budget for additional energy analysis.

The Strategic Profit Model

The final section of this chapter uses the Strategic Profit Model to illustrate the connection between an industrial organization's energy choices and its business performance. This exercise helps to coordinate key business managers and investors that would otherwise resist energy efficiency—by providing answers to the perennial question, "what's in it for me?"

Let's start with the obvious: an industrial enterprise exists to make money. Investors create the enterprise by providing capital equity to finance the assets that will produce goods for sale. If all goes well, the sale of these goods produces revenues in excess of the enterprise's capital investment and operating expenses. The financial efficiency of this process is captured in the "return on equity" concept. This concept is useful to investors that have many choices of where to invest their capital. Return on equity (ROE) is a *relative* measure that allows investors to compare the attractiveness of two or more investment alternatives.

ROE can be simply expressed as the ratio of new wealth created (variously described as *earnings* or *net income*) to the value of the investors' equity:

$$\text{ROE} = \frac{\text{Net Operating Income}}{\text{Equity}} \qquad \textit{(Eq 6-12)}$$

The ROE equation yields a percentage based on a ratio of dollar figures taken directly from consolidated financial statements. "Net operating income" is an entry on an income statement, while "equity" is recorded on a balance sheet. An ROE of 20 percent, for example, is the result of $2 million in annual income being produced by assets that are capitalized by $10 million in equity.

The elegant simplicity of the ROE concept is not without its detractors. During the 1920s, financial analysts with DuPont Corporation noted that ROE was a static measure that failed to provide insight on the business dynamics behind its simple ratio. They wanted more than a financial snapshot; they wanted a meaningful measure of financial *productivity*.

The analogy of *functional* productivity may help to explain the Dupont staff's vision for *financial* productivity. Let's say you own a very small, simple trucking business. You have one truck and one driver. You'd like to measure the functional productivity of the business. At the end of the day, you ask your driver what he accomplished. "I drove 70 miles per hour," he says. That's not enough, so you ask for more. "I drove for ten hours," he says. So now you have a little something: the driver covered 700 miles. But this doesn't describe what was accomplished. You find that he carried two tons of cargo. So now you have a basic productivity measure of *1,400 ton-miles*. This basic measure of productivity for this example is:

$$\text{Productivity} = \text{Speed} \times \text{Duration} \times \text{Capacity} \qquad \textit{(Eq. 6-13)}$$

In this example, *speed* provides an instantaneous measure of the velocity at which the task was undertaken. *Duration* describes the impact of speed over time, otherwise known as distance. But distance alone means nothing without a measure of *capacity* carried. Taken together, these variables indicate productivity. The separate variables allow a manager to isolate, evaluate, and address the individual factors that contribute to overall productivity.

The DuPont analysts anticipated a similar way to breakdown financial performance measures. Specifically, the question was: What drives the organization's overall profitability? The Dupont Formula, also referred to as the *Strategic Profit Model,* traces the linkages between departments and profitability. The model developed from this beginning:

$$\text{Return on Assets} = \frac{\text{Net Operating Income}}{\text{Revenues}} \times \frac{\text{Revenues}}{\text{Average Assets}}$$

(*Eq. 6-14*)

Return on assets (ROA) proves to be an intermediate result, the same way that *distance* was only a partial measure of productivity for the hypothetical trucking company. As an intermediate metric, ROA is relevant to the operational performance of assets currently in place. It may be most useful for comparing the management performance between two or more facilities owned by the same company. In other words, it describes how well one facility performs relative to all other facilities within a company in using assets to create wealth. But by definition, this statistic focuses on assets without distinguishing between the debt and equity that capitalize those assets. As such, it fails to describe shareholder returns, i.e., returns specific to equity by itself. That oversight is addressed by the basic Strategic Profit Model:

$$\text{Return on Equity} = \frac{\text{Net Operating Income}}{\text{Revenues}} \times \frac{\text{Revenues}}{\text{Average Assets}} \times \frac{\text{Average Assets}}{\text{Average Equity}}$$

(*Eq. 6-15*)

where:

- all values are from the same accounting period
- net operating income is the remainder of total revenues after operating expenses are deducted
- revenues are total sales receipts
- average assets is the total value of the organization's assets, expressed as the mean value of two consecutive end-of-year balances
- average equity is represents the value of all shareholder investment, expressed as the mean value of two consecutive end-of-year balances

A mathematically inclined reader will notice that in Equation 6-15, the terms "revenue" and "average assets" would both cancel out, leaving *net operating income* divided by *average equity* as the result. But that would miss the point—the intention here is to isolate the separate contributions to business performance. The formula is simplified to isolate these contributions as follows:

$$\text{Return on Equity} = \text{Operating Margin} \times \text{Asset Turnover} \times \text{Financial Leverage}$$

(*Eq. 6-16*)

Operating margin (net income divided by revenues). This is a relative financial measure of the organization's efficiency in converting raw materials and other inputs (not including investor capital) into revenue. To the investor, an industrial facility is a money-making machine: one dollar's worth of inputs goes in one end, and some amount in excess of one dollar should come out the other end. An acceptable net operating margin reflects the minimization of costs as well as the appropriateness of the price at which the product is sold. There is a connection to energy in that production expenses vary directly with the reduction of energy waste. There's an indirect relationship in that energy optimization efforts usually have spill-over benefits in the form of increased productivity, reduced scrap rates, reduced emissions and safety liabilities, and the ability to bring new products to market that demonstrate a minimized environmental footprint.

Asset turnover (revenues divided by average assets). Again, this is a relative financial measure. If the industrial enterprise is a "pipeline" for generating wealth, then asset turnover is a measure of that pipeline's capacity. Simply put, this metric measures how much work is being produced. Asset turnover is particularly useful for comparing the productivity of different facilities that make similar products. For example, say that one facility employs $100 million in assets to generate an annual production worth $300 million (asset turnover = 3.0). This compares favorably to a facility with $150 million in assets with annual production worth $350 million (asset turnover = 2.3).

Financial leverage (average assets divided by average equity). This metric describes the degree to which an organization uses borrowed wealth (or *debt*) to supplement the equity supplied by its investors. In other words, it's a measure of how much the business entity relies on other people's money to underwrite its assets. This concern usually rests with a chief financial officer and has no immediate bearing on day-to-day operating decisions. But financial leverage is relevant to energy optimization to the extent that debt financing may be used to pay for energy waste remediation or to finance the start-up of product lines that will be promoted for their environmentally friendly attributes, derived in part from the energy-efficient way in which they were manufactured.

The basic Strategic Profit Model (Equation 6-15) addresses pre-tax return on equity. A more advanced model provides a post-tax measure by incorporating tax and debt consequences. By pulling line items from consolidated financial statements for a specific accounting period, it looks like this:

$$ROE_{post\text{-}tax} = \frac{\text{NOPAT}}{\text{Pre-Tax Profit}} \times \frac{\text{Profit}}{\text{EBIT}} \times \frac{\text{Pre-Tax EBIT}}{\text{Revenues Assets}} \times \frac{\text{Revenues}}{\text{Average Equity}} \times \frac{\text{Average Assets}}{\text{Average}}$$

(Eq. 6-17)

where the variables are as defined above with Equation 6-15 PLUS:

- *EBIT* is "earnings before interest and taxes," and is synonymous with "net operating income" as used in Equation 6-15
- *Pre-tax profit* is the result of extracting interest costs from EBIT
- *NOPAT* is "net operating profit after taxes," or the remainder after subtracting taxes from pre-tax profit

In its full articulation (Equation 6-17), the Strategic Profit Model describes returns to investors after taxes. Equation 6-17 is also helpful in capturing the consequences of any tax benefits that accrue to the corporation. More to the point, the collection of any energy-related tax benefits can be neatly connected to shareholder results by using this formula. For example, investment in a renewable energy application may result in a tax credit. That credit directly reduces the difference between pre- and post-tax profit measures. Mathematically, the impact is apparent in Equation 6-17: an increased value for the ratio NOPAT to pre-tax profit is a proportional increase to post-tax return on equity.

PUTTNG THE PIECES TOGETHER

Let's restate Equation 6-17 by replacing the component ratios with their equivalent concepts:

$$\text{Return on Equity}_{post\text{-}tax} = \text{Tax Burden} \times \text{Interest Burden} \times \text{Operating Margin} \times \text{Asset Turnover} \times \text{Financial Leverage}$$

(Eq. 6-16)

Stated this way, the Strategic Profit Model is a tool for industrial organizations to "connect the dots" between staff accountabilities and returns to shareholders. The following lists strategic contributions to return on equity, a description of contributing activities, and an indication of who performs those activities:

OUTCOME: Minimize Tax Burden

- Chief engineers and finance officers:
  - Pursue tax incentives for investing in assets that improve energy and business productivity.

OUTCOME: Minimize Expenses to Increase Operating Margin

- Machine operators:
  - Shut down machinery when not in use.

- Office staff:
  - Shut down computers and other office equipment when not in use.

- Facilities manager:
  - Make facility-wide use of motion detectors, screen savers, timers and other controls to reduce energy waste.

- Maintenance staff:
  - Repair leaks in steam, compressed air, duct work, and other in-house utilities.
  - Optimize fuel-air mixtures for combustion.
  - Maintain adequate insulation on mechanical systems and in the building envelope.
  - Minimize friction in motors and motor drives.
  - Match motor horsepower to loads.
  - Ensure proper equipment start-up sequencing to avoid peak demand spikes and to prioritize the use of the most efficient assets.

- Production schedulers and operations managers:

— Schedule workloads to take advantage of time-of-use electricity tariffs.
— Coordinate production calendars with scheduled maintenance calendars.

- Procurement managers:
  — Develop an energy procurement strategy that minimizes purchase price for a given level of aversion to market risk.
  — Implement life-cycle cost criteria for procurement of energy-related hardware.

- Chief engineers:
  — Facilitate the utilization of energy consumption data that tracks usage to points of accountability throughout the facility.

- General managers:
  — Ensure that the costs and benefits of energy improvements are shared across department lines. For example, avoid situations where Dept. A pays for an improvement, but all the benefits accrue to Dept. B (and Dept. A is penalized for an extraordinary expense).

OUTCOME: Increase Revenues to Improve Operating Margin and Asset Turnover

- Product developers and marketers:
  — Develop products and / or services that can be marketed as environmentally friendly, thanks to the company's demonstrated reduction in energy waste.
  — Join supply chains organized exclusively for bringing environmentally friendly products to market.

- Chief finance officers:
  — Establish and prepare to sell carbon credits associated with the reduction of fossil fuel emissions as that market emerges.

- Chief engineers:
  - — Where markets permit, and if the company has the capacity, sell surplus electricity generated from onsite power plants.

OUTCOME: Increase Asset Values to Improve Financial Leverage

- Chief financial officers:
  - — Optimize depreciation values. Cost segregation is a hybrid engineering/accounting effort that enables this result. The purpose is to improve cash flow by reclassifying energy-related assets for tax purposes from a 39-year depreciation category to a 15- or 7-year category.

- Chief engineers, chief financial officers:
  - — Asset values increase as the risk of holding them declines. Therefore, efficient mechanical and production systems reduce add value to a facility by reducing operational risk. Investors and asset managers appreciate transparency of the controls and protocol for shielding a facility from potential liabilities. That transparency comes in part from the installation of energy metering, monitoring, and verification (MM&V) systems that provide managers with a real-time pulse of operations. With the appropriate MM&V system in place, staff are empowered to optimize energy performance. Overall plant value should increase as waste reduction contributes to asset turnover—in other words, allowing existing assets to work harder and produce more value. Additionally, MM&V systems contribute to the management of safety and emissions liabilities.

ADDRESSING APPARENT CONTRADICTIONS IN THE MODEL

Depreciation vs. cash flow. If an asset has already been placed in service, then with all else being equal, it is advantageous to accelerate its depreciation. This effectively increases near- to medium-term cash flows to the business. Annual cash flows are improved by an amount equal to income taxes that are avoided. But here lies the one contradiction: if depreciation is accelerated, then annual expenses are increased. This is a good thing in that it reduces tax liabilities (and improves cash flow). However, this is bad from the Strategic Profit Model's perspective. Increased depreciation means increased operating expenses, which reduces net operating income. In other words, this decreases the model's *operating margin* term. To counter this, a solid corporate energy plan would (1) seek a cost segregation analysis that identifies opportunities for accelerated depreciation; (2) use the cash flow derived from this depreciation exercise to invest in energy-related assets that provide tax credits or deductions, and (3) reap the operating expense savings that the new energy investments provide. This clearly requires some asset management collaboration among the organization's finance and engineering leadership.

Asset turnover vs. debt burden. The issue here is using debt to improve returns on shareholder equity (ROE). On one hand, borrowing increases financial leverage, which is good for ROE. But if that debt is used to capitalize additional assets, this decreases asset turnover (a bad outcome, from the shareholders' perspective). However, investors are made whole again if sufficient new revenues are generated by the new assets, therefore offsetting the decrement in asset turnover. The business strategy here can be to generate new revenues through the production of "green" and environmentally friendly product lines. A green product is made possible in part by an energy-efficient manufacturing process. Strategic planning would allow increased borrowing specifically to finance the capacity for producing "green" product lines. Meanwhile, the revenues from that production improve the operating margin while boosting asset turnover.

IMPLICATIONS AND CONCULSIONS

Truly efficient use of industrial energy requires a coordination of priorities within facilities and across layers of management. This feat is made difficult when decision-makers throughout the organization fail to see the connection between energy choices and bottom-line business performance. The antidote to this dilemma may be to refocus on money, and in particular, shareholder equity. The Strategic Profit Model, presented here, has been used for years by Wall Street for analyzing the performance of publicly traded companies. It can and should be adopted to promote energy efficiency to business leaders.

The Strategic Profit Model will coordinate engineering, operations, and finance decisions needed to maximize energy-efficiency investments. Not only that, the model is a blueprint for connecting those decisions to the primary business purpose of maximizing returns on shareholder equity. In effect, this framework counters the skeptics of energy efficiency by answering the question "What's in it for me?"

For a variety of reasons, manufacturing organizations fail to maximize return on equity. This is especially evident when competition among departments leads to the failure to spend a dime that would actually cause a whole dollar to be saved. The Strategic Profit Model helps to reconnect decision-makers with the true determinants of business performance. It does this by breaking down the return on shareholder equity into components that can be linked directly to discrete functions within the manufacturing organization.

Industrial energy efficiency can be harnessed to shareholders' benefit if the true "money" impacts are made clear to all stakeholders. The Strategic Profit Model is a framework for fostering the internal collaboration that industrial organizations need to ensure that energy improvements contribute to business performance.

Endnotes

1. To get really technical about it, investments should be amortized on a monthly basis. The annual expense is then the sum of 12 monthly amortized values.
2. http://www.ase.org/section/topic/industry/corporate/cemcases. Accessed March 2, 2009.
3. Thanks to Stan Walerczyk of Lighting Wizards for inspiring much of this discussion.

Chapter 7

Lessons and Outcomes

"The only risk of failure is promotion."
—Scott Adams
(1957-)
Author and creator of the "Dilbert" comic strip

Adams created "The Dilbert Principle," which asserts that companies systematically promote their least-competent employees to middle-management positions to get them out of harm's way. While many academics dispute this concept, a text with the same title remained on the *New York Times* bestseller list for 43 weeks.

Energy Lessons: The *Columbia* Disaster

On February 1, 2003, the NASA space shuttle *Columbia* was attempting to complete its twenty-eighth mission when it blew up during re-entry. The events leading up to this tragedy, as well as the recovery efforts in its aftermath, generated many lessons that are potentially of great value to industrial facility managers. In no way does this discussion mean to trivialize that event. In fact, because lives were lost, we owe it to those people to learn as much as possible from their experience.

It is not fair to say that mismanagement of facilities will lead to spectacular failures and loss of life (although industrial accidents do claim lives every year). More to the point, the mechanical integrity of many facilities is less than optimal, which absolutely costs money in terms of wasted fuel and productivity. The organizational causes of energy losses and accidents are strikingly similar to NASA's shuttle experience. Mechanically, there are only a couple of similarities between space shuttle operations and, for example, an industrial steam system: these are large pieces of machinery that generate a great deal of heat and force. The similarities are more pronounced in terms of management priorities, procedures, communication, data interpretation, and professional culture. Investigation into the *Columbia's* demise describes catastrophic management system failures. [1]

1. Clues to failure were available well in advance of the catastrophe, drawing attention to the shuttle's insulating tiles. Concerns with the integrity of these tiles date back to the shuttle's initial delivery in 1979. Because the tiles were a source of ongoing but minor concern, decision-makers were apparently lulled into a sense of complacency—it had yet to cause real problems, so why intervene? Think now about deferred maintenance in an industrial steam system. Is it appropriate to ignore water hammer by saying "it's always been like that?"

2. Perceptions and priorities were divided along professional lines in NASA. On one hand, technical staff focused on data, measurement, and verification, while program managers dealt with budget cutbacks, expense savings, and deadlines. The goals and teamwork between these two very different professional cultures were less than perfect. Many industrial energy improvements—especially those with zero-cost—are behavioral in nature. They may be the result of better coordination between utility and process staff. A lot of dollars can be saved just by thoughtfully planning the sequence in which assets are powered up.

3. Failure is often the result of a series of incidents, not just one. A chain of events may involve technology, communication, data interpretation, and the structure of accountability. Analysis of *Columbia's* failure was not limited to assembling

pieces of the stricken craft. It also involved an audit of email communications among staff. In countless manufacturing plants, equipment operators and maintenance staff never see the fuel bills that procurement staff process every month. If the data in those documents are not shared, then clues to operating anomalies and run-away costs remain hidden.

Let's shift now to the recovery efforts in the wake of the shuttle's crash. This was an effort that covered several states and involved everything from U2 spy planes to scuba divers and sniffing dogs. What was remarkable about the recovery effort was the volume of material it retrieved. Experts said that at best, 15 percent of the structure would be recovered. Through April 22, 2003, teams had in fact recovered almost 40 percent of *Columbia's* unfueled weight.

About 130 federal, state, and local agencies had to collaborate to make this happen. Usually, agency collaboration across jurisdictions are a recipe for confusion, red tape, and turf battles. This was largely avoided in the case of *Columbia's* recovery. What made this successful? And what are the lessons for industrial energy management?

There was clear and singular "ownership" of the process. All jurisdictional authority coalesced around NASA's lead. The lesson here for industry: an energy champion is usually vital to the success of energy management efforts. This is an individual with knowledge and authority to act. The "champion" is the visionary, coach, and arbitrator who keeps everything on track.

Clear and simple goals facilitated jurisdictional coordination. Look now at the manufacturing plant where utility, process, financial, logistical, and other managers must be on the same page. Business provides its own rallying cries: Return on investment. Building shareholder wealth. Global competition. To be effective, these managers have to know how their respective areas contribute to a central goal, and then focus their teams accordingly. Shared goals precipitate trust, which then yields better communication.

Skilled workforces bring the highest value. The *Columbia* recovery effort engaged "the best of the best" from each contributing jurisdiction. This task was not passed off to junior staff or pursued intermittently as people found free time away from their normal duties. As in any industrial operation, the value of motivation and accountability are underscored by this lesson.

A final thought regards sustainability of effort. The *Columbia* space shuttle disaster is an episode—a singular event that captured the dedication of staff involved in its closure. A tragedy so visible and of such magnitude naturally evoked focus on the part of recovery teams. Quite simply, there was a vast but finite acreage to cover, and they could declare victory when they covered it all. Manufacturing is not quite the same. Plants will operate year in and year out, and managers never have a true "finish line." Instead, they have a fluid business environment, and with that challenge also comes the opportunity to periodically adjust the plant's focus in achieving its goals. Therein lie the dynamics of motivation and lasting business success.

The Evolution of Sustainable Business

Sustainability describes principles that minimize negative environmental and social impacts, both now and in the future. Businesses are increasingly adopting sustainability principles, and their reasons for doing so are continually evolving. Sustainability principles are not directly compelled by law. Meanwhile, a prescription for attaining sustainable business remains somewhat elusive. ISO 14000 may offer the most rigorously developed guidelines. [2] ASHRAE Standard 100-2006, entitled Energy Conservation in Existing Buildings, attempts to prescribe sustainability principles for a more narrowly defined audience. [3]

"Sustainability" emerged as a business buzzword in the 1990s. The first business leaders to embrace the concept may have done so primarily for boosting their companies' public images. If so, then their investments in waste recycling and resource conservation may have been perceived as a way to grow their companies' value, as reflected in the goodwill line item on their balance sheets.

In 2002, the Sarbanes-Oxley Act was signed into U.S. federal law in response to recent examples of egregious corporate fraud. [4] Stated broadly, the Act attempts to ensure higher standards of corporate responsibility, especially regarding financial and accounting liabilities. Since environmental performance can have enormous financial implications, corporate sustainability programs emerged as a tool for offsetting the risk of failing to meet regulatory scrutiny. In other words, by investing in sustainability initiatives, corporations insured themselves against regulatory non-compliance.

Fast-forward to today. Popular awareness of climate change is influencing consumers' purchasing behavior. A growing number of consumers demand products and services that offer a reduced "environmental footprint." Wal-Mart advances the sustainability concept by challenging its suppliers to wring as much waste as possible from their manufacturing and distribution efforts. Eaton Corporation, a diversified manufacturer of auto and aircraft components, recently joined the Green Suppliers' Network, [5] so that its suppliers could coordinate efforts to "improve processes, increase energy efficiency, implement cost-saving opportunities and optimize use of required resources to eliminate waste." For suppliers, then, investment in sustainability programs becomes the cost of gaining access to markets... or perhaps the cost of simply staying in business.

Sustainability and the Triple Bottom Line

The "triple bottom line" concept is gathering momentum especially among publicly traded companies. In broad terms, it describes any corporate effort to focus on more than just financial results. We all understand the bottom line to mean "money," but for a variety of reasons, businesses are forced to recognize their environmental and social impacts as well. The sustainable business model is one that optimizes financial, environmental, and social performance all at once.

As you might suspect, there are plenty of business arguments about the merits of focusing on any outcome other than the creation of wealth. Nevertheless, businesses are increasingly scrutinized by consumers, shareholders, boards of directors, and other observers who challenge business operations that impose negative social or environmental impacts. Complaints against corporations encompass anything from unfair labor practices through the environmental impact of industrial waste handling. When a business "does good by doing no bad," it is minimizing its own costs while offsetting the risk of business lost due to adverse publicity. In the end, the triple bottom line still comes down to money.

Forward-thinking corporations are protecting themselves by demonstrating their social and environmental stewardship. Many companies simply offer the public a sustainable business vision statement. But at some point, words must be followed by action. The challenge to corporations is to pursue sustainability in a way that provides an economic return on the time and money that they invest in this effort.

Energy cost control is high on the list of sustainable business opportunities. Why? Because of all the measures a company can take to demonstrate its social and environmental credentials, energy improvements are best able to pay for themselves—and in the end, create more income. When a business pursues energy improvements, it will:

1. **Improve the "money" bottom line**. Energy consumption is reduced as waste is eliminated. This allows a company to buy less energy. Those savings translate, dollar for dollar, into operating income. At the very least, those savings can be used to generate investment income. Another option is to

reinvest the savings in the business in ways that create new revenue opportunities.

2. **Improve the "environmental" bottom line**. As energy waste is eliminated, the emissions from fossil fuels such as coal, oil, and natural gas are reduced proportionately. While these fuels are used in many industrial processes, they are also used for space and water heating. Even if your business is not industrial, or if does not use fossil fuels, it still uses electricity. Electricity is generated from a mix of fuel sources, including fossil fuels (notably coal). By turning on a light fixture or other electric appliance, a business is ultimately causing the combustion of fossil fuels. Carbon and other emissions from fossil fuels are the focus of environmental concerns. So even though your company is not in the business of generating electricity, it still has a direct, causal link to the amount of fossil fuel emissions released during the generation of electricity.

3. **Improve the "social" bottom line**. Anyone who uses electricity places demands on the national power grid; anyone who avoids energy waste reduces stress on that grid. The U.S. infrastructure for power generation and distribution is operating at capacity, and much of that capacity is in need of updating. The investment needed to meet future power demand is enormous. [6] Business and industry represent a large share of power demand (and the potential for waste reduction). Reducing waste helps to keep investment in electricity infrastructure to the minimum that is needed—and no more. The alternative is to live with energy waste, which means inflating investment in power capacity so that we can feed that waste. Anyone who doesn't generate electricity in his own backyard is forced to pay for investment in the national power grid. Cutting energy waste means not only savings on your energy bill, but ensuring that society's investment capital is not wasted on power capacity that can

be avoided. How's that for social responsibility?

Returns to the triple bottom line—economics, society, and the environment—await corporations that adopt a sustainable business model. Energy improvements can be the vanguard of that effort.

The Hunters and Farmers of Energy Management

Companies approach energy management with one of two basic strategies," a colleague once said. "One is that of the hunter, and the other, the farmer." That comment deserves its place as the epilogue of this text.

Before the advent of towns and market centers, frontiersmen were self-sufficient in providing not only daily sustenance, but (if times were good) some surplus commodities for the market. The same opportunities await manufacturers that proactively manage their energy resources—or in other words, by "hunting and farming" wealth from their own facilities.

The farmer stakes out a fixed territory and produces value by systematically sowing, tending, and harvesting valuable crops. The farmer's discipline of daily chores, patiently applied year after year, allow him to reap wealth from his acreage. Aside from some weather-related risk, the farmer can look forward to a predictable yield of commodities. The farmer may produce only one or two crop varieties per season, but the volume is enough not only to feed the farmer's family, but also to sell for cash income.

The hunter roams freely about the land in search of game. His task is opportunistic—relying on chance and skill to secure a small volume of meat and pelts. The hunter works hard for his bounty, but the goods return a much higher price per unit of mass than do the farmer's. The hunter's effort returns value very quickly, but the hunter shoulders a sizeable risk of failure for his time commitment. It is not unusual for a team of hunters to pool their talents when stalking their game.

So we would expect the frontier head-of-household to put some effort into both hunting AND farming, balancing his time wisely between the two tasks in a way that reflected the risks and rewards inherent in each activity. A clever individual could ensure the harvest of staple grains for his family with a surplus to generate cash. At the same time, the effort applied to hunting would bring meat for the dinner table as well as pelts that might bring in some extra income.

Now, let's apply this thinking to industrial energy management.

Energy is to the factory as fertile land is to the farmer. Through the distribution of electricity, steam, and compressed air, energy can potentially "fertilize" every square foot of space. This energy can either be harnessed to make products, or it can dissipate through waste. Remember that in either case, the plant pays for that energy.

The "farmer," in today's factory, harvests value from existing plant assets. He ensures that fuel, air, and water inputs are monitored and adjusted as needed. Leaks and losses from steam, air, water, and other distribution systems are minimized. Operating benchmarks indicate the optimal level of energy consumption per unit of production, while periodic data snap-shots indicate when systems stray from those benchmarks. Like the farmer who methodically plows each row of land, the industrial energy manager monitors each layer of energy utilization data. He develops a protocol for reacting to anomalies in the data. The industrial energy "farmer" needs to know the cost-benefit of taking remedial action. Benchmarks and operating data are his ledger—they are crucial for establishing guidelines for taking action. They are also evidence of the value he has saved. In other words, documentation of energy flows will demonstrate the total value saved (and the wealth it will generate) as well as the value of avoided waste.

The modern industrial analog to the "hunter" is the plant engineer who seeks singular pieces of technology, strategically chosen to improve operating effectiveness. Through consultation and research, the plant engineer scopes out new and improved applications that pay for themselves through the savings and extra productivity that they provide. He lobbies his corporate directors in the capital budgeting process. There's an elevated risk-reward aspect with the selection of strategic projects, but the successful engineer is skilled at both technical analysis and in presenting his proposals to corporate officers, to explain "what's in it for them." It is not unusual for the engineer to pool efforts with others who can help to secure his "game." In this instance, the engineer's allies are the technical assistance teams located at universities, professional societies, utilities, and consulting firms.

Industrial plant managers today are on the frontier of a challenging future. We know from history that frontiersmen survived by diversifying their modes of livelihood, and by teaming their skills and efforts with others. While some manufacturers will falter, others will thrive, especially if they harvest their resources wisely. Energy is the "fertilizer" of industry. Information is the ploughshare. Truly competitive manufacturers will enlist both "farmers" and "hunters" to reap wealth from their energy use.

Endnotes

1. http://caib.nasa.gov/ Accessed March 2, 2009.
2. http://www.14000.org/. Accessed March 2, 2009.
3. http://www.ashrae.org/pressroom/detail/15418. Accessed March 2, 2009.
4. http://frwebgate.access.gpo.gov/cgi-bin/getdoc.cgi?dbname=107_cong_bills&docid=f:h3763enr.tst.pdf. Accessed March 2, 2009.
5. http://www.greensuppliers.gov. Accessed March 2, 2009.
6. See http://www.eia.doe.gov/oiaf/aeo/index.html. Accessed March 13, 2008.

Appendix I

The New World of Energy Procurement

Volatile energy markets and the deregulation of gas and electric utilities are forcing industrial energy consumers to adopt energy procurement strategies. Business consumers seek protection from energy price spikes that can destroy earnings and upset budget performance. Risk is inherent in the way energy is both purchased and consumed, but organizational accountabilities for price control are usually far more stringent than they are for waste.

Management of price risk remains a necessary part of energy cost control. [1] Consumers want to be shielded not just from high prices, but also from volatile price movements that complicate fiscal and budget planning. In a deregulated energy market, consider the potential for price movement between the time when a consumer presents an offer to buy and the time when the transaction is fulfilled. "Price risk" describes the degree of market volatility between those two points in time. The techniques used to manage financial risks, like investments in stock and bonds, are directly transferable to energy procurement. The key concept here is hedging, or structuring a transaction for the express purpose of neutralizing the potential for lost value.

Hedging involves the assignment of risk. In other words, for a transaction with some lag time between contract ratification and actual fulfillment, who will absorb the risk of market price fluctuation—the buyer or the seller? The consumer who wants 100 percent certainty of energy expense must pay the supplier a premium for a contract to receive a commodity at a fixed price on a specified date. Conversely, the consumer who accepts the risk of market fluctuation evades that premium by purchasing indexed contracts, i.e., contracts with prices set to reflect the natural ebb and flow of the particular market. Many consumers blend their consumption with a combination of fixed and indexed contracts. A related hedging tool is the "option," which gives the bearer the right, but not the obligation, to procure energy at a fixed price within a prescribed period of time.

A fixed-price contract makes sense for the consumer who anticipates any upward movement in the future price of energy. A contract can lock in a chosen price for a specific quantity of energy. The consumer who purchases 100 percent of its energy this way "assumes a fully-hedged position," or is shielded against the potential for higher market prices in the future. But by the same token, such contracts—as obligations to buy at a fixed price—prevent the consumer from enjoying market price dips. The opposite end of the risk spectrum is the "fully-indexed position." This means, for an energy consumer, making all purchases at the market price that prevails at the time of order fulfillment.

In reaction to energy price spikes in the wake of the 2005 hurricane season, many industrial energy consumers aggressively hedged their consumption through 2006. This means they purchased fixed-price contracts that anticipated continued upward movement in the market. However, energy prices dropped during 2006. In effect, the hedged consumers ended up paying higher prices—at least during 2006—than the prevailing market. Does this experience mean that hedging is a bad strategy? To answer, think of it this way: chances are that you paid for homeowner's insurance in 2006, but you did not need to make a claim against your policy. That doesn't mean that your insurance expenditure was a waste. Think of energy procurement hedging as a form of insurance—in this case, against the risk of dramatic price spikes.

An important point to remember: by consistently assuming a fully-hedged position, the consumer pays more for a commodity in the long run. This is because of the premium that is paid for the surety of a fixed-price contract. In contrast, by assuming a fully-indexed position, the consumer absorbs the risk of market volatility. This entails enduring the occasional price spike, but it also allows the consumer to enjoy dips in prices and to consistently avoid paying risk premiums.

Procurement strategies can only be a partial solution to high energy costs. Hedging only stabilizes price risk. Consumers seeking to proactively and consistently reduce energy expenditures must reduce their energy waste. This requires a business plan that employs technologies, procedures, and behaviors for continuous energy improvement.

Endnotes

1. Thanks to Skip Trimble and Tucker Davis of South River Consulting for their patient tutoring on this subject.

Appendix II

Electricity Deregulation Explained for the Industrial Consumer

BACKGROUND

Most people have heard about electricity deregulation, but very few understand what it means. In brief, deregulation began as an attempt to restore fairness to electricity markets that have, over the last couple of decades, outgrown the 1930s business model upon which they are based. In practice, deregulation has been a fitful process for consumers, utilities, and lawmakers that are caught in the middle. Consumers have come to expect the market for electricity (or "power") to be simple, fair, and offer low prices. Unfortunately, consumers can only obtain any two of those virtues, usually at the expense of the third. The good news is that consumers in deregulated power markets can choose which two virtues they wish to optimize. As will be explained in this appendix, access to electricity can be:

- simple & cheap, while compromising fairness;
- simple and fair, which generally precludes cheap prices; or
- cheap & fair, which is not a simple market for its participants.

Regulation is the legacy of utility business models from years past, which in turn reflected the technology of the times. In the U.S., prior to 1935, electric companies could and did clutter cities and streets with competing sets of distribution wires and ancillary equipment. Federal legislation passed that year would enable the regionalized-monopoly business model, so that only one utility company's infrastructure served a defined geographic area. In this format, a utility offered the simplicity of one supplier with one price schedule. In return for a monopoly franchise, utility businesses were closely regulated by state-chartered public utility commissions. Commissioners were tasked with representing the consumers' interest by ensuring that the rate for electricity covered the utility's operating costs while providing a fair return to the utility's investors—and no more. From the 1930s through deregulation's take-off in the late 20th century, electric utility companies provided steady if unspectacular financial returns. Large institutions have come to rely on utility-issued equity as a key part of their investment strategies.

By the 1990s, a number of forces began to challenge the classic electric utility business model. Perhaps the most important forces were the escalating cost of fuels needed to generate power; the growing variety of power generation technologies; the advance of information systems that permit faster, more precise, and more detailed interpretation of electricity generation costs, and the increased willingness of large-quantity consumers to relocate their facilities in search of better electric rates, or to simply generate their own power onsite. The economic tension created by these forces demanded resolution.

Immediately, questions of fairness come to mind. Why should the large consumer be forced to accept the local utility's electric rate when it can buy cheaper power in another location? Why should smaller customers subsidize the cost of serving larger ones? Why should the steady investment returns to utility investors be interrupted? If these returns are interrupted, utilities will find it a lot harder to attract investors. The only cure for that is to raise the rate of return on utility investments. In other words, the cost of capital for utilities would escalate, which in turn raises the cost to produce (and the price to buy) electricity. As electricity prices go up, common citizens begin to petition lawmakers for protection. Average citizens (and lawmakers) do not understand the cost-price relationship for producing electricity; many people assume that electricity prices can be determined by the stroke of a pen, regardless of its actual cost of production. Does it not seem unfair to have elected officials decide the price of electricity? Exactly how are they supposed to do that?

The solution to this dilemma is to open electricity markets to competition. Power generation can be accomplished by competing suppliers. Meanwhile, the local distribution and service portions of electricity provision remain the same—it's still not practical to have competing sets of wires lining the street. By allowing competition among generators, big consumers can get access to lower electricity prices without relocating. But if you open the market to the big consumers, you must open it for all the small businesses and homes, too.

HOW DOES DEREGULATION WORK?

Welcome to the 21st century. Electricity purchase and consumption is a multi-part task. Keep in mind that electricity consumption is paid for in three parts: (1) the energy commodity itself, in units of kilowatt-hours, or kWh; (2) the capacity for delivering the energy (if electricity is the "beer," then the capacity is the "mug"); and (3) the ancillary services that ensure the safety and reliability of the distribution system itself. The electricity market constantly adjusts the output of power generators in response to the demands placed on them. A customer's total expenditure for electricity reflects not just the price and volume of these components, but also the time at which they are consumed.

The traditional regulated utility "bundled" together the provision of electricity—a commodity—with the cost of distributing power and servicing the related infrastructure. In contrast, the deregulated market allows customers to shop for commodities in the open market. In this market, the local utility retains the responsibility for distribution and service, but acts as a middleman between power generators and the consumer, making wholesale electricity purchases that are in turn delivered to retail consumers.

Regulators recognize that most residential and small business electricity consumers don't wish to be saddled with the many choices presented by deregulation. Therefore, the traditional utility company remains the "provider of last resort," which allows consumers to maintain the simplicity of a traditional utility relationship. However, the utility may not always provide the lowest of prices, especially when its procurement strategy is to make one or two large annual wholesale power purchases in a market that experiences price fluctuations in 15-minute intervals.

The open market for electricity is notoriously volatile for a variety of reasons. The demand for electricity varies with the weather and the time of day. Power generation facilities are not all created equal: some are more expensive to run than others. The cheapest-to-run generators tend to operate the most. The more expensive units add to power supply as demand increases, as it does on hot weekday afternoons when air conditioning is at a premium. Predicting electricity demand is about as reliable as a meteorologist's ability to predict the weather. Coordinating power generation capacity is an imperfect task, and as a result, demand and supply imbalances create price spikes and volatility.

Whereas the price of an office chair stays constant, 24 hours a day, seven days a week for a long period of time, electricity prices vary by the 15-minute interval. You can store an inventory of chairs in a warehouse until they are sold, but you can't do the same with significant amounts of electricity.

PARTICIPATING IN DEREGULATED ELECTRICITY MARKETS

The consumers who exercise choice in electricity markets pursue a trade-off among simplicity, price, and fairness. A consumer usually can obtain any two of these merits at the expense of the third. These are the options:

"Simple and cheap" at the expense of fairness. Electric utility service requires a lot of capital investment in distribution infrastructure. Utilities hate to lose a large customer because the capacity to serve that customer becomes idle. There's no revenue to offset the ongoing carrying costs of this idle capacity—unless the cost is passed on to remaining utility customers. By the way, these other customers as a community may already be suffering because that big facility has moved elsewhere for cheaper electricity rates, taking with it the employees who supported local businesses. The simple and allegedly cheap solution would be to redistribute the utility's cost of serving the large ex-customer among the many remaining small consumers, so that each bears a small burden. After a series of these redistributions, the remaining customers will ask if their growing cost burden is indeed fair.

"Simple and fair" is not always cheap. Would it not be fair and simple to have electricity prices reflect the cost of its production? This sounds good until we recall that the cost of power production is tremendously volatile even within the course of a single day. A price spike at the wrong time could ruin the budgets of businesses and households alike. In these instances, electricity prices pegged to the cost of production expose consumers to upside price risks.

"Cheap and fair" is not simple. Today's volatile electricity prices can be managed with sophisticated portfolio management techniques borrowed directly from the financial sector. This involves paying a premium for contracts to receive electricity at predetermined prices, terms, and quantities in advance of a specified date. Note that these contracts do not provide "cheap" prices. Instead, the contract provides a reliable price that ostensibly reduces the consumer's exposure to future, unpredictable price spikes. The consumer pays the supplier a premium (an increment of value over and above the electricity commodity itself) for the privilege of having a reliable price. Portfolio strategies allow a consumer to cover some or all of their anticipated electricity needs for months or even years in advance. The portfolio manager follows daily power markets for opportunities to make contract purchases. This may sound complicated because it is, especially for residential and small business consumers. Larger power consumers either develop a staff professional to perform these tasks, or enlist a consultant to do the same.

The turmoil that follows in the wake of electricity deregulation has been attributable to a lack of preparation—on the part of suppliers and consumers. Deregulation has been subject to political compromise, particularly when electricity price caps were imposed for an interim period to "phase in" deregulated markets. Maryland experienced exceptional market turmoil when price caps set in 1999 were lifted as planned in 2006—exposing consumers to the prevailing cost of electricity. The timing was terrible, as it coincided with the aftermath of the 2005 hurricane season, which had caused extensive damage to the natural gas supply infrastructure upon which electricity production depends. Residential consumers faced a potential 70 percent rate hike for electricity. In response to popular outcry, legislators intervened in the rate-making process to "force-fit" low prices, which effectively usurped the administrative power of Maryland's utility regulators. The "low price" solution, by the way, was one that simply deferred consumer costs over time.

Not surprisingly, there is popular support in deregulated markets such as Maryland for restoring traditional regulation to the electricity market. This means putting power generation and distribution under the care of a local monopoly.

NEAR-TERM OUTLOOK

Proceeding with electricity deregulation is like squeezing toothpaste from a tube: you can only go forward, and you can't put the toothpaste back in the tube. This is because deregulation required utilities to sell off the assets that generate electricity and to keep the assets used to distribute power to customers. To go back to the old business model, utilities would be forced to build or buy back power generating plants. This initiative would have enormous costs. Who would pay for these plants?

- If private investors foot the bill, then the utility needs to guarantee those investors a rate of return that is at least as attractive as other investment opportunities. Returns to the investor would have to be reflected in the rate for electricity sold by these new plants.

- If electricity consumers pay the utility to bring these plants online, this also means raising electricity rates.

- What about the "government?" This means raising the money through taxes. This money still comes from consumers' pockets.

Add to this dilemma the potential for market friction between neighboring states. The delivered price for electricity could be very different in neighboring states simply because of each state's regulatory approach. States that attempt to re-regulate electricity may impose costs that result in electricity prices higher than those in a neighboring state without that burden. The desire for price arbitrage across state lines would benefit consumers while possibly harming utility investors. This renews questions about fairness. Investors may look to their lawmakers for relief.

As the first decade of the 21st century draws to a close, only a handful of states have implemented electricity deregulation. Their progress has been stalled by memories of California's 2003 deregulation debacle, which was neither well designed nor well executed. Some states, like Virginia, are indeed taking steps to move backward in the deregulatory process. But for reasons discussed here, the outlook is for more deregulation, not less. Currently, the advantage falls to the informed consumer who is prepared to exercise the choices—however complex they may be—offered by today's deregulated utility markets.

Index

For Product Safety Concerns and Information please contact our EU representative GPSR@taylorandfrancis.com Taylor & Francis Verlag GmbH, Kaufingerstraße 24, 80331 München, Germany

Printed and bound by CPI Group (UK) Ltd, Croydon, CR0 4YY
08/05/2025
01864425-0001

Managing Energy From the Top Down

Connecting Industrial Energy Efficiency to Business Performance

Christopher Russell, C.E.M.

n a world with increasingly constrained resources, energy captures top management attention. Using language that is accessible to as many readers as possible his book explains the connection between energy and business performance Corporate leaders, production managers, machine operators, policy advocates and technology providers will all learn how day-to-day choices relate to the risks and rewards of energy use. Concise, to-the-point chapters explain how energy s invested, preserved, and ultimately positioned to create wealth. Hard-nosed business leaders should appreciate the section with examples that show a strong financial case for energy improvements, including the save-or-buy criterion, the economic penalty for "doing nothing," the break-even cost, and the budget for supporting design and analysis work. *Managing Energy From the Top Down* is written with the goal of making the challenges and opportunities of energy use accessible to all readers that have a stake in industrial competitiveness.

Christopher Russell, C.E.M., has enjoyed a career that features a consistent focus on business analysis, solution development, and project justification. Since 1999, he has become a nationally-recognized expert in the design and implementation of corporate energy cost control. Russell has documented and evaluated energy management practices at dozens of facilities, and has advised corporations, utilities, trade associations, and government agencies in the planning and promotion of industrial energy programs. He is also in high demand as a writer consultant, and keynote speaker at industry conferences. He has been an energy columnist for *Maintenance Technology, Insulation Outlook*, and *Chemical Processing* magazines. He is recognized by the Association of Energy Engineers both as a Certified Energy Manager and a Certified Energy Procurement Specialist. Russell joined the Board of Directors of the Fuel Fund of Maryland in 2006 He is on the Advisory Board for the Texas A&M Industrial Energy Technology Conference, and is a Board Member for the Management Standard for Energy (MSE) 2000 development project. He holds an MBA and a Master of Arts from the University of Maryland, and a Bachelor of Arts from McGill University in Montreal Canada. His energy management blog may be found at http://energypathfinder.blogspot.com. He resides with his wife and daughter in Baltimore, Maryland.

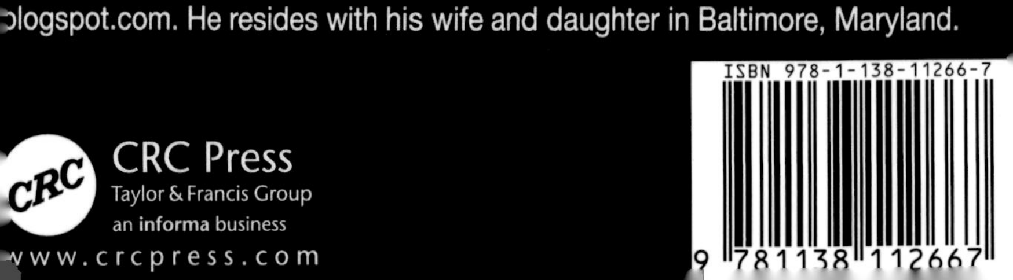
ISBN 978-1-138-11266-7
9 781138 112667

CRC Press
Taylor & Francis Group
an informa business
www.crcpress.com